Geomatics in Tsunami

NIPA® GENX ELECTRONIC RESOURCES & SOLUTIONS P. LTD.
New Delhi-110 034

Geomatics in Tsunami

Edited for the **NRDMS** Division, Department of Science and Technology
by

SM. Ramasamy
Centre for Remote Sensing, School of Geosciences,
Bharathidasan University, Tiruchirappalli

C.J. Kumanan
Centre for Remote Sensing, School of Geosciences,
Bharathidasan University, Tiruchirappalli

Brig. R. Sivakumar
NRDMS Division, Department of Science and Technology, New Delhi

and

Bhoop Singh
NRDMS Division, Department of Science and Technology, New Delhi

NRDMS Division,
DST, New Delhi

Centre for Remote Sensing,
Bharathidasan University,
Tiruchirappalli

NIPA® GENX ELECTRONIC RESOURCES & SOLUTIONS P. LTD.
New Delhi-110 034

NIPA® GENX ELECTRONIC
RESOURCES & SOLUTIONS P. LTD.

101,103, Vikas Surya Plaza, CU Block
L.S.C. Market, Pitam Pura, New Delhi-110 034
Ph : +91 11 27341616, 27341717, 27341718
E-mail: newindiapublishingagency@gmail.com
web: www.nipabooks.com

For customer assistance, please contact
Phone: + 91-11-27 34 17 17
Fax: + 91-11- 27 34 16 16
E-Mail: feedbacks@nipabooks.com

ISBN: 978-81-19215-01-0

Composed and Designed by NIPA®.

कपिल सिब्बल
KAPIL SIBAL

राज्य मंत्री (स्वतंत्र प्रभार)
विज्ञान एवं प्रौद्योगिकी तथा महासागर विकास
भारत सरकार, नई दिल्ली
MINISTER OF STATE (INDEPENDENT CHARGE) FOR
SCIENCE & TECHNOLOGY AND OCEAN DEVELOPMENT
GOVERNMENT OF INDIA,
NEW DELHI

FOREWORD

The Tsunami that occurred on 26th December 2004 has sent shock waves literally amongst the scientific community in India as we had very little knowledge about this phenomenon. Immediately after the tsunami, a number of initiatives have been taken up to understand the mechanisms which triggered the tsunamis and possible early warning signals that could be generated. However, not much work has been done for assessment of damage caused both to flora and fauna, soils, buildings and other infrastructure. The teams that were sent out by the Department of Science and Technology soon after the occurrence of tsunami suggested carrying out damage assessment mapping all around the affected coastal Peninsula in the form of damage assessment mapping along with additional attribute information for use of GIS tools.

I am happy to see the outcome of hard work done by the DST team all along the coast of Tamil Nadu, Andhra Pradesh and Andaman and Nicobar islands. The report not only contains the scientific issues but also provide basic inputs for evolving guidelines for mitigating the hardships by tsunamis and all such other disaster management and mitigation activities. I congratulate the Secretary, Department of Science and Technology and the team involved in this activity for bringing out this compilation well in time.

(KAPIL SIBAL)

प्रो॰ व. सु. राममूर्ति
सचिव
PROFESSOR V. S. RAMAMURTHY
SECRETARY

भारत सरकार
विज्ञान और प्रौद्योगिकी मंत्रालय
विज्ञान और प्रौद्योगिकी विभाग
टेक्नालॉजी भवन, नया महरोली मार्ग, नई दिल्ली-110 016

GOVERNMENT OF INDIA
MINISTRY OF SCIENCE & TECHNOLOGY
DEPARTMENT OF SCIENCE & TECHNOLOGY
Technology Bhavan, New Mehrauli Road, New Delhi-110 016

MESSAGE

Soon after the occurrence of tsunami on 26th December, 2004, Hon'ble Minister for Science & Technology and Ocean Development had instituted a small team to visit the affected areas. The team has recommended mapping of the damaged areas for recording the possible damages occurred to the buildings, infrastructure, flora and fauna. The associated issues such as desalination of drinking water, mapping for GIS and also analyzing the deposits for metals and minerals besides recording the erosion have been adequately dealt. This study also helped us to understand the complex relationships between the communities and government departments. I am happy to note that this entire work was completed within a short span of ten months and this Volume is now for readers to comment upon.

(**V. S. Ramamurthy**)

Dr. C. THANGAMUTHU
Vice - Chancellor

BHARATHIDASAN UNIVERSITY
PALKALAIPERUR
TIRUCHIRAPPALLI - 620 024
TAMIL NADU

FOREWORD

The "Tsunami" that devastated most of the maritime nations of south Asia on December 26, 2004 was altogether a new phenomenon and hence warranted deeper scientific explanations and solutions to facilitate the planners and administrators to protect the coastal people, their properties and the valuable public infrastructure. Besides the tsunami warning system (TWS) contemplated, the Department of Science and Technology, Government of India, New Delhi, the apex body on S&T issues in the country has evolved an "All India Coordinated Mission Mode Programme" to have first hand information on the disaster. The programme has been executed by six agencies in the country (including the Centre for Remote Sensing, Bharathidasan University, Tiruchirappalli, Tamil Nadu). The tsunami disasters and risks are considered on a larger geospatial matrix, providing scientific guidelines for the planners to work upon possible mitigation strategies.

The Centre for Remote Sensing, Bharathidasan University, in addition to participating in the mission as one of the lead agencies has also brought out a special volume " GEOMATICS IN TSUNAMI" as text book of international standard. The editors Dr.SM.Ramasamy and Dr. C.J. Kumanan from the Centre deserve to be appreciated as the Centre always takes up very challenging and innovative assignments in the frontier areas of remote sensing applications. Besides the distinguished faculty of Bharathidasan University, Brig.R.Sivakuamr, Advisor and Dr. Bhoop Singh, Director, NRDMS, DST, New Delhi have also joined hands in bringing out this volume comprising quite significant scientific articles emerging from the experience of coordinated programme. These articles have been written by very eminent scholars from different perspectives. I am aware that this volume will provide not only a knowledge addition but also will be source of good baseline data for the planners and administrators involved in tsunami mitigation strategies. The book published at this juncture is quite topical and relevant and, I am sure, it would be well received by the scholars and policy makers in the field.

C. THANGAMUTHU

Preface

The Tsunami which has been an unknown terminology to the common man and least studied phenomenon by the scientists and technocrats of Indian ocean countries, has struck most of the coastal nations of the Indian Ocean on 26th December 2004. As this ferocious natural disaster swallowed over two lakhs people and shattered properties worth of several million rupees, the Government of India has taken up speedy measures to understand the phenomenon and to provide short term protective and long term preventive measures for tsunami. The Department of Science and Technology, through NRDMS Division, has envisioned an all India coordinated programme for assessing the tsunami damages in respect of natural, physical and human resources using Remote Sensing (Indian Remote sensing Satellite data of the Pre-Tsunami period and Digital Orthophoto and other satellite data of Post-Tsunami period) and to finally create GIS databases on the disasters, on 1:8,000 scale so as to provide baseline information for the planners and also to evolve possible futuristic strategies for tsunami mitigation from the zones and extent of damages. This study was executed through specialized centers in Remote Sensing, GIS and Geosciences in the country for the affected coastal areas of Andaman - Nicobar, Andhra Pradesh, Tamil Nadu and Pondicherry. Through these studies, GIS data bases were generated on the damages to natural resources (Coastal landforms, land use / land cover, soil in terms of salinity and alkalinity, modifications in aquifer etc.), damages to the buildings, communication and transportation, Agriculture, Irrigation, Human resources etc.

In addition to the results of the above studies, few more scientific papers invited from eminent scholars of diverse disciplines constitute this most illuminated and illustrative volume on text book standards entitled "**GEOMATICS IN TSUNAMI**".

The scientific articles embedded in this book address all the major issues viz: historical perspective, wave propagation, geomatics based damage assessment-interface dynamics between the tidal waves and the coastal landforms-the mitigative strategies evolved therefrom, impacts on the ecosystem, tsunami triggered societal issues, citizen's perspective, strategic roles played by Govt. and NGO's during 2004 tsunami etc., in five well defined sections.

The first two editors gratefully acknowledges Prof C. Thangamuthu, Vice- Chancellor, Bharathidasan University for his overwhelming support in undertaking this assignment. They also record their great appreciation to the research, technical and administrative team of them from Centre for Remote Sensing *viz*: Dr. K. Palanivel, Mr.N.Nagappan, Mr.J.Saravanavel, Mr.M.Muthukumar, Mr.T.Chandrasekar, Mr.T.Sreenivasulu, Mr.Sk.Md. Sartaj Basha, Mr.J.Muralitharan, Mr. V. Agastheeswaran, Mr. D. Sankar, Mr. Steny Thomas, Miss. Mary Vasanthakumary, Mrs. Shanmuga valli, Mr.S. Jesuraj, Mr. Suresh kumar, Mr. Govindaraj, Mr. Arokya Yesu Das and others, for their untiring effort and committed involvement in bringing out this book. Mr. R. Selvakumar, Research Associate and Mr.S. Gunasekaran, Research Scholar need special appreciation for ably coordinating the work. The editors record their deep sense of gratitude to Dr. V. S. Ramamurthy, Secretary, Department of Science and Technology for his leadership and to other members of NRDMS division for their support.

The staff of M/S New India Publishing Agency, New Delhi need a special appreciation and acknowledgement for publishing this book with self styled elegance. The editors also convey their thanks to all the eminent scholars who have contributed the scientific articles.

December, 2005
Tiruchirappalli, India

SM. Ramasamy
C. J. Kumanan
Brig. R. Sivakumar
Bhoop Singh

Contributors

V. Agastheeswaran, Centre for Remote Sensing, Bharathidasan University, Tiruchirappalli, Tamil Nadu– 620023.

Annie George, NGO, Co-ordination centre, 31, Mahalakshmi Nagar, South Palpannai cheri, Nagapattinam, Tamil Nadu - 611001.

R. Arthur James, School of Marine Sciences, Bharathidasan University, Tiruchirappalli, Tamil Nadu– 620024.

Arun Kumar, Department of Earth Sciences, Manipur University, Imphal, Manipur – 795001.

Bhoop Singh, NRDMS Division, Department of Science & Technology, Technology Bhavan, New Mehrauli Road, New Delhi – 110016.

N. Chandrasekar, Centre for GeoTechnology, School of Technology, Manonmaniam Sundaranar University, Tirunelvel, Tamil Nadu – 627012.

R. Chandrasekaran, Department of Disaster Management, SASTRA Deemed University, Thanjavur, Tamil Nadu – 613402.

R. K. Chingkhei, Department of Earth Sciences, Manipur University, Imphal, Manipur – 795001.

J. Dajkumar Sahayam, Centre for GeoTechnology, School of Technology, Manonmaniam Sundaranar University, Tirunelveli, Tamil Nadu – 627012.

T. Dhanapal , Principal Scientific Officer, TNSCS & T, DOT Campus, Sardar Patel Marg, Chennai, Tamil Nadu – 600025.

M. Dharma Raj, Tamil Nadu & Pondicherry, Director, Geo-Spatial Data Center, Survey of India, Chennai, Tamil Nadu – 600032.

S. Kalaivani, Centre for Environmental Studies, Anna University, Chennai, Tamil Nadu - 600025.

M.Kamaraj, Institute of Remote Sensing, Anna University, Chennai, Tamil Nadu - 600025.

L. Kannan, Centre of Advanced Study in Marine Biology, Annamalai University, Parangipettai, Tamil Nadu – 608502.

C.J. Kumanan, Centre for Remote Sensing, Bharathidasan University, Tiruchirappalli, Tamil Nadu - 620023.

J. Loveson Immanuel , Centre for GeoTechnology, School of Technology, Manonmaniam Sundaranar University, Tirunelveli, Tamil Nadu - 627012.

R. Mahesh, Department of Disaster Management, SASTRA Deemed University, Thanjavur, Tamil Nadu - 613402.

K. Mruthyunjaya Reddy, Andhra Pradesh State Remote Sensing Application Centre, 2nd Floor, Directorate of Economics and Statistics (DES) Campus, Khairatabad, Hyderabad, Andhra Pradesh - 500004.

T.S. Murty, Department of Civil Engineering, University of Ottawa, Ottawa, Canada.

K. Muthumeena, Department of Disaster Management, SASTRA Deemed University, Thanjavur, Tamil Nadu - 613402.

Nageswara Rao, Andhra Pradesh State Remote Sensing Application Centre, 2nd Floor, Directorate of Economics and Statistics(DES) Campus, Khairatabad, Hyderabad, Andhra Pradesh - 500004.

RM. Narayanan, Institute of Remote Sensing, Anna University, Chennai, Tamil Nadu - 600025.

Niladri Gupta, Geoinformatics & Remote Sensing Cell, West Bengal State Council of Science & Technology(DST), Govt. of West Bengal, Bikash Bhavan, Salt Lake, Kolkata - 700091.

I. Nistor, Department of Civil Engineering, University of Ottawa, Ottawa, Canada.

N. Nirupama, School of Administrative Studies, York University, Toronto, Canada.

K. Palanivel, Centre for Remote Sensing, Bharathidasan University, Tiruchirappalli, Tamil Nadu - 620023.

Parthasarathi Chakrabarti, Geoinformatics & Remote Sensing Cell, West Bengal State Council of Science & Technology(DST), Govt. of West Bengal, Bikash Bhavan, Salt Lake, Kolkata - 700091.

V.Raghu , Andhra Pradesh State Remote Sensing Application Centre, 2nd Floor, Directorate of Economics and Statistics(DES) Campus, Khairatabad, Hyderabad, Andhra Pradesh - 500004.

A. Rajarajeswari, Department of Disaster Management, SASTRA Deemed University, Thanjavur, Tamil Nadu - 613402.

G.V. Rajamanickam, Department of Disaster Management, SASTRA Deemed University, Thanjavur, Tamil Nadu – 613402.

M. Rajamanickam, Centre for GeoTechnology, School of Technology, Manonmaniam Sundaranar University, Tirunelveli, Tamil Nadu – 627012.

V. Rajeshkumar, Department of Disaster Management, SASTRA Deemed University, Thanjavur, Tamil Nadu – 613402.

A. Rajendran, School of Marine Sciences, Bharathidasan University, Tiruchirappalli, Tamil Nadu - 620024.

N.Rajendran, Centre for History, Bharathidasan University, Tiruchirappalli, Tamil Nadu - 620024.

M. Ramalingam, Institute of Remote Sensing, Anna University, Chennai, Tamil Nadu - 600025.

SM. Ramasamy, Centre for Remote Sensing, Bharathidasan University, Tiruchirappalli, Tamil Nadu – 620023.

A.D. Rao, Department of Civil Engineering, University of Ottawa, Ottawa, Canada.

D. Sankar, Centre for Remote Sensing, Bharathidasan University, Tiruchirappalli, Tamil Nadu – 620023.

R.Santhanam, Special Commissioner and Commissioner of Revenue Administration and State Relief Commissioner, Chennai, Tamil Nadu - 600009.

E. Saranathan, Department of Disaster Management, SASTRA Deemed University, Thanjavur, Tamil Nadu – 613402.

J. Saravanavel, Centre for Remote Sensing, Bharathidasan University, Tiruchirappalli, Tamil Nadu – 620023.

D. Satheesh Herbert Singh, Centre for GeoTechnology, School of Technology, Manonmaniam Sundaranar University, Tirunelveli, Tamil Nadu – 627012.

V. Sathiskumar, Department of Disaster Management, SASTRA Deemed University, Thanjavur, Tamil Nadu – 613402.

Sanjoy Nag, Geoinformatics & Remote Sensing Cell, West Bengal State Council of Science & Technology(DST), Govt. of West Bengal, Bikash Bhavan, Salt Lake, Kolkata – 700091.

R. Selvakumar, Centre for Remote Sensing, Bharathidasan University, Tiruchirappalli, Tamil Nadu – 620023.

V.Selvam, M.S.Swaminathan Research Foundation, III Cross Street, Taramani, Chennai, Tamil Nadu - 600113.

Steny Thomas, Centre for Remote Sensing, Bharathidasan University, Tiruchirappalli, Tamil Nadu - 620023.

Brig. R. Siva Kumar, Head, NRDMS Division, Department of Science & Technology, New Delhi - 110016.

D. Soosai Manickaraj, Department of Disaster Management, SASTRA Deemed University, Thanjavur, Tamil Nadu - 613402.

R. Sridhar, Centre of Advanced Study in Marine Biology, Annamalai University, Parangipettai, Tamil Nadu - 608502.

C. Sriramkumar, Department of Disaster Management, SASTRA Deemed University, Thanjavur, Tamil Nadu - 613402.

A.V. Subba Rao, Andhra Pradesh State Remote Sensing Application Centre, 2nd Floor, Directorate of Economics and Statistics(DES) Campus, Khairatabad, Hyderabad, Andhra Pradesh - 500004.

Subrata B. Dutta, Geoinformatics & Remote Sensing Cell, West Bengal State Council of Science & Technology(DST), Govt. of West Bengal, Bikash Bhavan, Salt Lake, Kolkata - 700091.

Sujatha E Ramani, Department of Disaster Management, SASTRA Deemed University, Thanjavur, Tamil Nadu - 613402.

Susmita Dasgupta, Geoinformatics & Remote Sensing Cell, West Bengal State Council of Science & Technology(DST), Govt. of West Bengal, Bikash Bhavan, Salt Lake, Kolkata - 700091.

T. Thangaradjou, Centre of Advanced Study in Marine Biology, Annamalai University, Parangipettai, Tamil Nadu - 608502.

Usha Natesan, Centre for Environmental Studies, Anna University, Chennai, Tamil Nadu - 600025.

C.Venkatachalam, Lecturer, Department of Sociology, Bharathidasan University, Tiruchirappalli, Tamil Nadu - 620024.

K.Vaithiyanathan, Institute of Remote Sensing, Anna University, Chennai, Tamil Nadu - 600025.

Contents

TSUNAMI –

HISTORY, MECHANISMS AND PHOTOGRAMMETRY

GEOMATICS IN TSUNAMI
S.M. Ramasamy, C.J. Kumanan, Brig. R. Sivakumar and Bhoop Singh (eds.), pp. 1-9
New India Publishing Agency, New Delhi (India)
E-mail : spjain_niph@rediffmail.com Web: www.bookfactoryindia.com

1

HISTORY OF TSUNAMI

N. Rajendran

Abstract

The paper deals with the historical perspective of the tsunamis, much widely discussed Kumari Kandam and Lemuria continent and the palaeo-time series changes on the past continents and oceans, disastrous tsunamis since 16th century and archeological evidences indicating shore line changes along Tamil Nadu coast.

Keywords: Ancient tsunamis in history, Kumari Kandam and Lemuria, Past shore line changes.

INTRODUCTION

A tsunami is a natural phenomenon consisting of a series of waves generated when water in a lake or the sea is rapidly displaced on a massive scale. Earthquakes, landslides, volcanic eruptions and large meteorite impacts all have the potential to generate a tsunami. The effects of a tsunami can range from unnoticeable phenomenon to large scale devastation.

The term "tsunami" is derived from the Japanese language meaning harbour ("tsu") and wave ("nami"). The term was coined by fishermen who returned to port to find the area surrounding the harbour devastated, although they had not been aware of any wave on the high seas. A tsunami is not a sub-surface event in the deep ocean; it simply has a much smaller amplitude (wave heights) offshore, and a very long wavelength (often hundreds of kilometers long), which is why they generally pass unnoticed at sea, forming only a passing "hump" in the ocean.

Tsunamis were historically referred to as tidal waves because when they approach land they take on the characteristics of a violent onrushing tide rather than the sort of cresting waves that are formed by wind action upon the ocean (with which people are more familiar). However, since they are not actually related to tides the term is considered misleading and its usage is discouraged by oceanographers.

CAUSES OF TSUNAMI

A tsunami can be generated by any disturbance that rapidly displaces a large mass of water, such as an earthquake, volcanic eruption, and landslide or meteorite impact. However, the most common cause is an undersea earthquake. An earthquake which is too small to create a tsunami by itself may trigger an undersea landslide which in turn quite capable of generating a tsunami.

Tsunamis can be generated when the sea floor abruptly deforms and vertically displaces the overlying water. Such large vertical movements of the earth's crust can occur at plate boundaries. Subduction earthquakes are particularly effective in generating tsunamis, and occur where denser oceanic plates slip under continental plates in a process known as subduction.

Sub-marine landslides; which are sometimes triggered by large earthquakes; as well as collapses of volcanic edifices, may also disturb the overlying water column as sediments and rocks slide down the slope and are redistributed across the sea floor. Similarly, a violent submarine volcanic eruption can uplift the water column and generate a tsunami.

Waves are formed as the displaced water mass moves under the influence of gravity to regain its equilibrium and radiates across the ocean like ripples on a pond.

In the 1950s it was discovered that larger tsunamis, than the previously believed ones, are possible through the landslides, explosive volcanic actions and impact events that rapidly displace large volumes of water, as energy from falling debris or expansion is transferred to the water into which the debris falls. Generally speaking, tsunamis generated from these mechanisms, unlike the ocean-wide tsunamis caused by some earthquakes, dissipate quickly and rarely affect coastlines distant from the source area due to the small area of sea affected. However they can give rise to much larger shock waves (solitons) locally, such as the Lituya Bay mega tsunami which produced a water wave estimated at 50 - 150 m and reached 524 m up local mountains.

CHARACTERISTICS OF TSUNAMI

Although often referred to as "tidal waves", a tsunami does not look like the popular impression of "a normal wave only much bigger". Instead it looks rather like an endlessly onrushing tide which forces its way around and through any obstacle. Most of the damage is caused by the huge mass of water behind the initial wave front, as the height of the sea keeps rising fast and floods powerfully into the coastal area. The sheer weight of water is enough to pulverise objects in its path, often reducing buildings to their foundations and scouring exposed ground to the bedrock. Large objects such as ships and boulders can be carried several miles inland before the tsunami subsides.

In open waters, tsunamis have extremely long periods (the time for the next wave takes to pass a point after the previous one), from minutes to hours, and long wavelengths of up to several hundred kilometers (compared to the typical wind-generated swell one sees at a beach, which might be spawned by a far away storm and rhythmically roll in, one wave after another, with a period of about 10 seconds and a wavelength of 150 m). The actual height of a tsunami wave in open water is often less than one meter. This is often practically unnoticeable to people on ships. The energy of a tsunami passes through the entire water column to the sea bed, unlike surface waves, which typically reach only down to a depth of 10 m or so.

The wave travels across the ocean at speeds from 500 to 1,000 km/h. As the wave approaches land, the sea shallows and the wave no longer travels as quickly, so it begins to 'pile-up'; the wave-front gets compressed becomes steeper and this gains an amplitude of 30 meters or more. The steepening process is analogous to the cracking of a tapered whip. As a wave goes down the whip from handle to tip, the same energy is deposited in the whip with less and less diameter, hence it moves violently towards the tip with gradual increase in amplitude.

Tsunamis occur most frequently in the Pacific Ocean, but are a global phenomenon; they are possible wherever large bodies of water are found - including inland lakes. Historically the earth has witnessed quit a few Tsunamis.

According to Greek sources, some where between 1650 BC and 1600 BC (still debated), the volcanic Greek island Santorini erupted, causing a 100 m to 150 m high tsunami that devastated the north coast of Crete, 70 km (45 miles) away, and would have certainly eliminated every timber of the Minoan fleet along Crete's northern shore. Santorini is regarded as the most likely source for Plato's literary parable of Atlantis, and is believed by some scientists to have informed Great Flood accounts which were eventually recorded in Jewish, Christian and Islamic texts.

TSUNAMI : THE HISTORICAL RECORD

Kumari Kandam and Lemuria

Kumari Kandam has often been compared and identified with Lemuria. According to Tamil Tradition, the Dravidians originally came from a submerged island Kumari Kandam in the south of India. The Epics *Shilappadikaram* and *Manimekhalai* describe the submerged city of "Puhar".

Also in Ayyavazhi mythology, the "*Akilattirattu Ammanai*" tells about a sunken land at about 152 miles either south or south-east to Kanyakumari, with 16008 streets. It also claims that the Center of origin of Human beings as the sunken land or the land surrounding it.

Lemuria is the name of a hypothetical "*Lost Land* " located in the Indian and Pacific Oceans. Its 19th century origins lie in attempts to account for discontinuities in biogeography. Lemuria has been rendered superfluous by modern understanding of plate tectonics.

Accounts of Lemuria differ according to the requirements of their contexts. However, all share a common belief that the continent existed in ancient times but sank beneath the ocean as a result of geological changes, often cataclysmic.

Geologists today regard sunken continents as physical impossibilities, given the isostatic equilibrium of continental plates floating on the thermoplastic mantle.

Scientific Origins

Though living modern Lemurs are found only in Madagascar and several surrounding islands, the biogeography of extinct lemurs extending from Pakistan to Malaya inspired the name Lemuria, which was coined in 1864 by the geologist Philip Sclater in an article "The Mammals of Madagascar" in *The Quarterly Journal of Science*. Puzzled by the presence of fossil lemurs in both Madagascar and India, but not in Africa nor the Middle East, Sclater proposed that Madagascar and India had once been part of a larger continent, which he named Lemuria for its lemurs.

Sclater's theory was hardly unusual for his time. The acceptance of Darwinism led scientists to seek to trace the diffusion of species from their points of evolutionary origin; prior to the acceptance of continental drift, scientists frequently postulated submerged land masses in order to account for populations of land-based species now separated by barriers of ocean. Many hypothetical submerged land bridges and continents were proposed during the 19th century, in order to account for the present distribution of species (biogeography).

As Lemuria gained some acceptance within the scientific community, it began to appear in the works of other scholars. Ernst Haeckel, a German Darwinian taxonomist, proposed Lemuria as an explanation for the absence of "missing link" fossil records. Locating the origins of the human species on this lost continent, he claimed the fossil record could not be found because it had sunk beneath the sea.

Other scientists hypothesized that Lemuria had extended across parts of the Pacific ocean, explaining distributions of species across Asia and the Americas.

In 1894, Frederick Spencer Oliver published *A Dweller on Two Planets*, which claimed that survivors from a sunken continent called Lemuria were living in or on

Mount Shasta in northern California. The Lemurians lived in a complex of tunnels beneath the mountain and occasionally were seen walking the surface dressed in white robes.

This belief has been repeated by such individuals as the cultist Guy Warren Ballard in the 1930s who established the I AM Foundation. It is also repeated by followers of the Ascended Masters and the Great White Brotherhood. This list includes such organizations as Bridge to Freedom, Summit Lighthouse, Church Universal and Triumphant, Temple of the Presence, and Hearts Center.

Fanciful archeologists, in the days before Mayan glyphs, who translated a section of the late Mayan period Madrid Codex, (that is sometimes called the Troano Codex) thought that they were able to interpret illustrations as "records" of a continent in the Pacific, destroyed by volcanic activity. Supposedly, a similar legend has been translated from unspecified "Sanskrit tablets" that describes a continent called Rutas.

The continent of Mu imagined by Augustus Le Plongeon (1826–1908) and James Churchward is possibly a permutation of ideas about what Lemuria might have been.

Lemuria's Reptilian Beings

In reptilian conspiracy literature, a sunken Pacific continent (usually styled as Lemuria or Mu) is sometimes posited as the homeland of a reptilian race of creatures, often identified with Dragons or Nagas. Various bits of myth and folklore are assembled in support, such as the Cambodian Naga traditions. Modern claims of Australian aborigines sighting "dinosaur-like" creatures are also often viewed as evidence.

The earliest attestation of such notions in modern literature seems to have occurred in the works of H.P. Blavatsky, notably in *The Secret Doctrine* (1888), where she writes of "Dragon-men" who once had a mighty civilization on a Lemurian continent, till their rampant use of black magic brought about the end of their civilization, and their continent sank. Blavatsky in turn claims to have gotten this information from *The Book of Dzyan*. However, many consider that Blavatsky invented the Book herself. Blavatsky believed that the terms "Dragon-men" or "Serpent-men" used to describe the Lemurian beings in the *Book of Dzyan* were symbolic, intended to symbolize their advanced knowledge and magical powers.

Another early occurrence of the idea seems to be in the Alley Oop (1932) comic-strip, where lands named Moo and Lem (adapted from Mu and Lemuria respectively) are presented as dinosaur-infested lands.

RECENT SHORE LINE CHANGES ALONG TAMIL NADU COASTS

Mahabalipuram

At Mahabalipuram, near Chennai, submerged ruins have been found in the ocean. There exists a popular belief in south India that the Shore Temple of Mahabalipuram is the last of a series of seven temples, six of which have believed to have been submerged. The members of a Scientific Exploration Society, U.K. explored this area. The archaeologists from NIO, trained in diving, and professional divers and underwater cameramen carried out underwater exploration during April 2002 and recorded evidence of ruins off Mahabalipuram.These investigations were carried out at 5 locations in the 5-8 m water depths, 500 to 700 m off shore temple. Investigations at each location have shown presence of the stone masonry construction, remains of walls, a big square rock cut, scattered square and rectangular stone blocks, big platform leading the steps to it amidst of the geological formations of the rocks that occur locally. Most of the structures are damaged and scattered in a vast area, providing substratum for barnacles, mussels and other sedentary organisms. The construction pattern and area is about 100x50m. The actual area covered by ruins may be much larger. The Pallava dynasty, ruling the area during the period, had constructed many such rock cut and structural temples in Mahabalipuram and Kanchipuram. A very initial time frame for the ruins is the same as those of the temples in the vicinity. Local belief also supports this observation.

The National Institute of Oceanography is processing the data collected, and are will working with the Archaeological Survey of India and eventually they will date the findings. As this was only a preliminary, reconnaissance survey substantial work is to be done and dates and context may vary based on such future findings.

The National Institute of Oceanography is studying the problem from various angles such as palaeo-shoreline and the geology in addition to the archaeological aspects. Full-scale investigations are underway to determine the role of sea level fluctuations, coastal erosion and neo-tectonic activities in effecting shoreline changes in the area during the recent past. The site has great potential for further exploration. Lay out plan of the structures is being determined as well as the reasons for its submergence. The findings are significant from the point of the view that while earlier there was merely folklore, now there exists "prima facie" evidence, and though a lot more remains to be done. If archaeological evidence proves these to be temple ruins and not natural phenomena, it will still be required to understand the cause of submergence of the structures and ensure that these are consistent with palaeo-oceanographic understanding of the area.

Poompuhar

Poompuhar known as Kaveripattinam is also one of the most notable ancient Chola ports and is located at the confluence of river Cauvery in the Bay of Bengal. *Manimekhalai* mentions that the Chola king had lost his son and in his grief he forgot to celebrate the annual Indra festival. This enraged the goddess who caused the city to be swallowed up by the sea. Though the reference is only to a supernatural incident, however, it may be taken as an echo of some actual sea erosion due to high tidal waves that lashed out on the city.

Onshore exploration near Chinnavanagiri, south of Poompuhar brought to light a ring well along with megalithic black and red wares. A few ring wells were also noticed near Vanagiri, Tranquebar and Poompuhar. A Sangam period brick structure was exposed near the present Kanagi statue. Another brick structure in L-shape was noticed in inter-tidal zone during low tide. These findings suggest that the shoreline is moving landward and a large part of ancient city of Poompuhar must have been submerged in the sea.

Offshore explorations opposite to mouth of river Cauvery brought to light stone structures. By airlift operation, a large number of pottery belonging to 4th-5th century AD (1600 – 1700 Y.B.P) were recovered. Similarly, in the north of Poompuhar significant pottery of historical period was found. These evidences indicate that a part of ancient city was submerged in the sea, the remains lying at about 8 m water depth. A mid-17th century map of Tranquebar shows a complete plan of the town along with the then shoreline. A careful study of the map suggests that Tranquebar town was well protected by a seaward fort wall and the Siva temple was sufficiently landward within the fort wall. It is estimated that during 17th century AD the shoreline was at least 50 m away from the fort wall and a temple was located around 250-300 m from the shoreline. This observation unequivocally suggests that the shoreline has transgressed about 300 m in the last 300 years infringing at an average rate of one meter per year. If the same trend had continued since the last 2000 years, it might be expected that the ancient city of Poompuhar must have been submerged in 5 to 10 m water depths off Poompuhar. About 3.5 km off Poompuhar a shipwreck was discovered at a water depth of 19 m, a major portion of which is buried under sediments. A metal detector survey suggested a maximum length of 50 m and width about 15 m. Airlift operation have shown that the ship was wooden hulled. The exploration of shipwreck brought to light a large number of lead ingots of three types: (i) boat shaped with an inscription of W:BLACKETT and a date 1791, (ii) A Roman letter 'D' with crown symbol and boat shaped and (iii) plain with rectangular shape. Besides the lead ingots, 2.1 m long iron cannon, gun-powder boxes and a copper rudder gudgeon were also noticed. Airlifting shipwreck suggests that wood and copper were extensively used for construction of the ship. The extent of wood boring also suggests that the wreck site was exposed for a long time. There are no records to ascertain the country of origin of the ship.

HISTORY OF TSUNAMI IN THE OTHER PARTS OF THE WORLD

A Massive Tsunami Struck, Portugal in 1755

Tens of thousands of Portuguese who survived the great 1755 Lisbon earthquake were killed by a tsunami which followed a half hour later. Many people fled to the waterfront, believing the area safe from fires and from falling debris from aftershocks. Before the great wall of water hit the harbour, waters retreated, revealing lost cargo and forgotten shipwrecks.

The earthquake, tsunami, and subsequent fires killed more than one third of Lisbon's pre-quake population of 275,000. Historical records of explorations by Vasco da Gama and Christopher Columbus were lost, and countless buildings were destroyed (including most examples of Portugal's Manueline architecture). The destruction of Lisbon sharply checked the colonial ambitions of the Portuguese Empire. Europeans of the 18th century struggled to understand the disaster within religious and rational belief systems. Philosophers of the Enlightenment, notably Voltaire, wrote about the event. The philosophical concept of the sublime, as described by philosopher Immanuel Kant in the "Observations on the Feeling of the Beautiful and Sublime", took inspiration in part from attempts to comprehend the enormity of the Lisbon quake and tsunami.

Many animals sensed danger and fled to higher ground before the water arrived. The Lisbon quake is the first documented case of such a phenomenon in Europe. The phenomenon was also noted in Sri Lanka in the 2004 Indian Ocean earthquake. Some scientists speculate that animals may have an ability to sense subsonic Rayleigh waves from an earthquake minutes or hours before a tsunami strikes shore.

Krakatoa Explosive Tsunami

The island volcano of Krakatoa in Indonesia exploded with devastating fury in 1883, blowing its underground magma chamber partly empty so that much overlying land and seabed collapsed into it. A series of large tsunami waves was generated from the explosion, some reaching a height of over 40 meters above sea level. Tsunami waves were observed throughout the Indian Ocean, the Pacific Ocean, the American West Coast, South America, and even as far away as the English Channel. On the facing coasts of Java and Sumatra the sea flood went many miles inland and caused such vast loss of life that one area was never resettled but went back to the jungle and is now the Ujung Kulon nature reserve.

Chilean tsunami

The Great Chilean Earthquake, at magnitude 9.5, the strongest earthquake ever recorded, off the coast of South Central Chile in 1960 generated one of the most destructive tsunamis of the 20th century. It spread across the entire Pacific Ocean, with waves measuring up to 25 meters high. When the tsunami hit Onagawa, Japan, almost 22 hours after the quake, the wave height was 3 m above high tide. The number of people killed by the earthquake and subsequent tsunami is estimated to be around 2,290.

Good Friday tsunami

An Earthquake, tsunami struck Alaska, British Columbia, California and coastal Pacific Northwest towns, killing 122 people after a strong earthquake scaling 9.2 in Richter scale on a Good Friday in 1964. The tsunami waves were up to 6 m tall, and killed 11 people as far away as Crescent City, California.

2004 - Indian Ocean tsunami

The magnitude 9.0 Indian Ocean Earthquake triggered a series of lethal tsunami on December 26, 2004 that killed more than 280,000 people (more than 220,000 in Indonesia alone), making it the deadliest tsunami in recorded history. The tsunami killed people over an area ranging from the immediate vicinity of the quake in Indonesia, Thailand and the north-western coast of Malaysia to thousands of kilometers away in India, Sri Lanka, the Maldives, and even as far as Somalia, Kenya and Tanzania in eastern Africa.

Unlike in the Pacific Ocean, there is no organized alert service covering the Indian Ocean. This is in part due to the absence of major tsunami events since 1883 (the Krakatoa eruption), which killed 36,000 people. In light of the 2004 Indian Ocean tsunami, UNESCO and other world bodies have called for a global tsunami monitoring system.

CONCLUSION

The glimpse of the historical records shows that there are definite evidences for marine transgression and regressions and land - ocean modifications, some of which could be tsunami. Hence detailed studies involving drilling, sediments sample collection, characterization dating etc., along the shores would unravel many vital questions about the past tsunamis and the time series analysis of which would give predictive equations on future tsunamis. The east coast of India, due to its well developed geomorphology is an excellent candidate for the study.

GEOMATICS IN TSUNAMI
S.M. Ramasamy, C.J. Kumanan, Brig. R. Sivakumar and Bhoop Singh (eds.), pp. 11-16
New India Publishing Agency, New Delhi (India)
E-mail : spjain_niph@rediffmail.com Web: www.bookfactoryindia.com

2

THE ROLE OF OSCILLATIONS OF THE FIRST AND SECOND CLASS IN TSUNAMI AROUND ANDAMAN AND NICOBAR ISLANDS

T.S. Murty, N. Nirupama, I. Nistor and A.D. Rao

Abstract

There are two distinct types of oscillations in global oceans, separated in their frequencies by the period of the pendulum day. One species is the gravity waves and the other is the rotational waves, associated with earth's rotation. Both these species can be found in tidal records around islands as well as near coastlines. These two types of modes are usually found in the tsunami records on tide gauges. The tide gauge records as well as visual descriptions of the water levels during and after the occurrence of a tsunami clearly show the presence of these oscillations. During the tsunami of 26th December 2004 in the Indian Ocean, media reports suggested high water levels persisted around the Andaman and Nicobar Islands for several days. These persistent high water levels can be explained by invoking the existence of trapped and partially leaky modes on the shelves surrounding these islands.

Keywords: Oscillations in global waves, Tsunami (2004), Andaman and Nicobar islands.

INTRODUCTION

The analysis of tidal observations revealed that there are two distinct types of oscillations in the oceans that are separated in their frequencies through the pendulum day. One species is the gravity waves, also referred to as Gravoids modes or Oscillations of the First Class (OFC). The second species is Oscillations of the Second Class (OSC) or rotational modes and is also called Elastoid-inertia oscillations. The first class has gravity explicit in their frequency equation, while for the second class, which owe their existence to earth's rotation, gravity does not play an important role in their frequencies. The importance of these two types of oscillations in the world oceans is known for a long time and especially in the tide gauge records near coastlines with wide continental shelves and also in tidal records around islands. The bathymetry of the wide shelf near a coastline or the bathymetry on the shelf surrounding an island can trap a significant amount of long gravity energy associated with such events as tsunamis and storm surges and retain it on the shelf for several days, if not weeks, until the trapped energy slowly leaks away out of the shelf. We suggest that trapped modes were the most probable cause of high water levels around Andaman and Nicobar Islands (Fig.1) for several days after the tsunami.

Waves oscillation models

In this paper we present some theory and some simple analytical calculation which shows that oscillations with periods of few minutes are possible. However, an exact comparison has to wait until a more sophisticated numerical model study

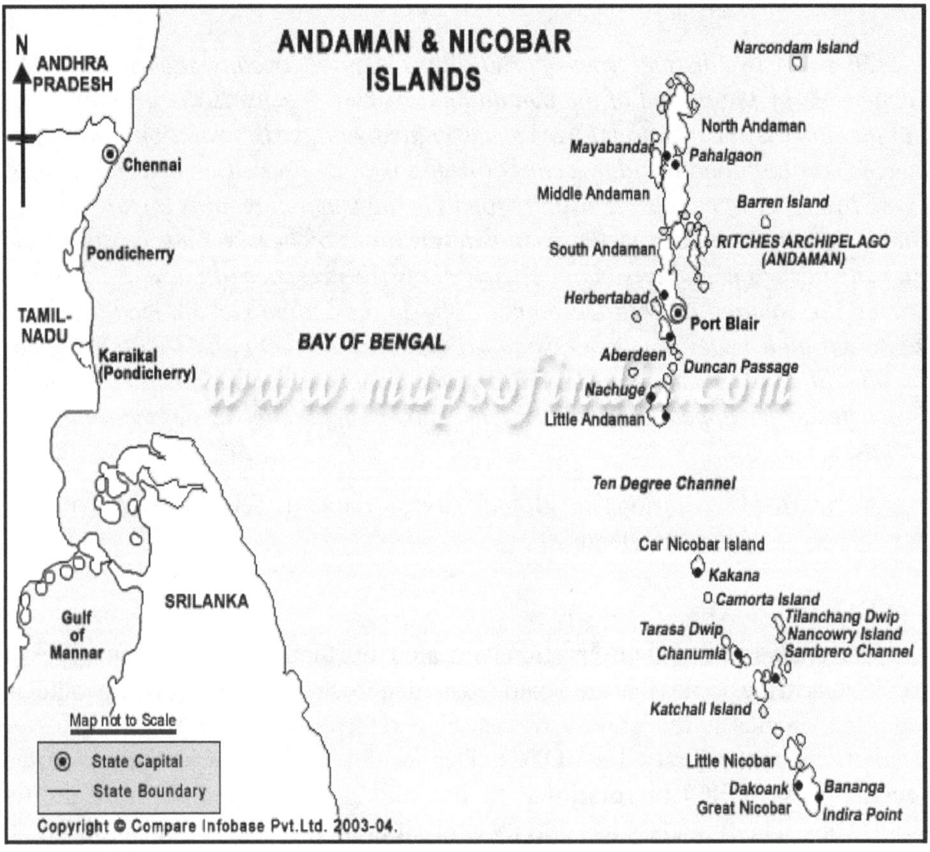

Figure 1: Detailed map of Andaman and Nicobar Islands region

is completed as well as a thorough analysis of all the available tide gauge records is done. But one result that has become obvious even in this preliminary study is that in the trapped modes around Andaman and Nicobar Islands following the tsunami, there were several oscillations with periods of the order of several minutes and we suggest that these are the Gravoid and Elastoid modes.

Trapped and Partially Leaky Modes

There have been several classical approaches in literature in the past. More recent work on trapped wave motion has been done by Liu et al (2003).

First-Class, Trapped-Wave Motions

Gravity waves generated in shallow water may get totally reflected while proceeding into deep water at the continental slope (Eckart 1950). Isacks et al (1951) used ray theory concepts to show that these waves could be captured from the open sea and guided along a long coastline by successive reflections between the shoreline and the offshore water. Trapped waves can be isolated from other motions (Munk et al 1964) and thus it can be concluded that trapped waves account for most low-frequency energy. The important fact that tsunamis approaching a coast can generate trapped waves, has been conclusively shown with reference to the Japanese coast by Hatori and Takahasi (1964). The theoretical discovery about these modes was first made by Stokes (1846). Munk et al (1956) found edge waves from water level records taken during the passage of storms (see also Freeman and Murty 1972 and Murty and Freeman 1973).

Trapping gravity wave energy (caused by guiding waves along the continental shelf) is somewhat analogous to other wave guides in geophysical problems (e.g. Love waves in the earth's crust and sound waves in the Sound Fixing and Ranging (SOFAR) channel in the oceans). Ursell (1951) showed that surface waves could be trapped along a submerged circular cylinder. Munk and Arthur (1952) used ray theory to show that gravity waves could be guided along underwater ridges. An important paper on trapped modes is Longuet-Higgins (1967) who showed that isolated islands might trap long waves that are guided by local bathymetry. He showed from theoretical calculations for an idealized circular seamount or sill, that there are some oscillations or modes trapped by the mount. He showed further that waves travelling from the open make them detectable in shallow-water wave records. Arthur (1951) showed that the effect of the local bathymetry in refracting waves around an island is an important factor in creating large waves on the sea shore.

Second-Class, Trapped-Wave Motions

The so-called oscillations of the second class (OSC) owe their existence to the earth's rotation. These waves also could get trapped like the gravity waves discussed earlier. Reid (1958), while studying the effect of Coriolis force on OFC, found another type of very low-frequency motions that propagate along (in the same direction as Kevin waves) and are almost confined to the shoreline, which he called quasi-geostrophic boundary waves. For these waves, the longshore component of the fluid flow and the offshore wave slope approximately balance each other.

Longuet-Higgins (1968) theoretically showed that for continental shelf waves to exist, the shallow-water region could have infinitesimal width. He showed that under uniform rotation, long waves could be trapped along a straight-line discontinuity separating two regions, each of uniform depth and the trapped waves were the OSC-type and propagated along the discontinuity only in the same direction of travel as Kevin waves in the deeper of the two regions. Rhines (1969a, b) considered an ocean on a β-plane where β is the variation of the Coriolis parameter *f,* with latitude. Although on the globe β varies with latitude, in the β-plane approximation it is treated as uniform. His study showed that the motions are guided along coherent and closely spaced contours of the ratio *f*/D, where D is the depth of the ocean.

The motions considered here under OSC have periods greater than a pendulum day, defined as the period of revolution of a Foucault's pendulum. This is given by:

$$\text{Pendulum day} = 2T_p = \frac{2\pi}{\Omega \sin\theta} \tag{11}$$

Where θ is the latitude and Ω is the angular velocity of earth's rotation. This expression shows that the pendulum day varies from place to place, depending on latitude. It is 2π/Ω at the pole and tends to reach ∞ at the equator. When the rotation Ω → 0 the OSC become steady currents.

Computation of the Periods of Trapped Modes

Yanuma and Tsuji (1998) gave the following relationship for edge wave modes trapped on the continental shelf:

$$\omega^2 = \frac{(2N+1)(2M+1).\pi.\alpha.g}{2L} \tag{29}$$

Where, α = slope of the shelf

ω = angular frequency of the edge wave mode

g = acceleration due to gravity = 9.8 m/sec^2
L = width of the shelf
N = number of the progressive edge wave mode
M = number of the standing edge wave mode
Period of the edge wave mode = T = $2\pi/\omega$ (30)

For standing edge wave modes, N = 0

The period of standing edge wave modes is given by:

$$T = \frac{2\pi\sqrt{2L}}{\sqrt{(2M+1)\,.\pi\,.\alpha\,.g}} \tag{31}$$

If we take L = 1 km and α = 1/1000, for the first standing edge wave mode, M = 0

$$T_{S_0} = \frac{2\pi\sqrt{2\times1000}}{\frac{3.14\times9.8}{1000}} \text{ seconds} = \frac{2\sqrt{2}\times3.14\times1000}{\sqrt{3.14\times9.8}\times60} \text{ minutes} = 26.7 \text{ minutes}$$

For the partially leaky modes, the progressive wave number N will be an integer higher than zero.

CONCLUSIONS

Theoretical arguments and simple analytical computations have been presented to show that most probably the reason for persistent high water levels around the Andaman and Nicobar Islands following the tsunami of 26th December 2004 in the Indian Ocean was the presence of the Gravoid and Elastoid modes of oscillation. These modes are oscillations due to trapping of the long gravity wave energy associated with the tsunami, on the shelf surrounding these islands. While an exact comparison cannot be made until the numerical model computations are completed, it is worthwhile noting that, the periods of oscillations, of the order of several minutes, in the persistent high water levels around Andaman and Nicobar Islands following the tsunami, that were reported in the media, do agree reasonably well with the results of the simple calculation presented here.

REFERENCES

Arthur, R. S. (1951) The Effects of Islands on Surface Waves. Bull. Scripps Inst. Oceanography, v. 6, pp. 1-26.

Eckart, C. (1950) The Ray-Particle Analogy. Journal of Marine Research, v. 9, pp. 139-144.

Freeman, N. G. and Murty, T.S. (1972) A Study of a Storm Surge on the Lake Huron. Proceeding of 15th Conference, Great Lakes Res. Int. Assoc. Great Lakes Res., Ann Arbor, Mich. pp. 565-582.

Hatori, T. and R. Takahasi. (1964) On the Iturup Tsunami of Oct. 13, 1963 as Observed Along the Coast of Japan. Bulletin of Earthquake Research Institute, v. 42, pp. 543-554.

Isacks, J. D., Williams, E. A. and Eckart, C. (1951) Total Reflection of Surface Waves by Deep Water. Transaction of American Geophysical Union, v. 32, pp.37-40.

Liu P. L.-F., Lynett, P. and Synolakis, C. E. (2003) Analytical Solution for Forced Long Waves on a Sloping Beach. Journal of Fluid Mechanics, v. 478, pp. 101-109.

Longuet-Higgins, M.S. (1967) On the Wave-Induced Difference in the Mean Sea Level Between the Two Sides of a Submerged Breakwater. Journal of Marine Research, v. 25, pp. 148-153.

Longuet-Higgins, M.S. (1968) On the Trapping of Waves along a discontinuity of depth in a rotating Ocean. Journal of Fluid Mechanics, v. 31, pp. 417-434.

Munk, W. H. and Arthur, R. S. (1952) Wave Intensity along a refracted ray, In: Proceeding of Symposium on Gravity Waves. U.S. Natl. Bur. Stand. Circ. v. 521, pp. 95-108.

Munk, W. H., Snodgrass, F. E. and Carrier, G. (1956) Edge Waves on the Continental Shelf. Science, v. 123, pp. 127-132.

Munk, W. H., Snodgrass, F. E. and Gilbert, F. (1964) Edge Waves on the Continental Shelf: An Experiment to Separate Trapped and Leaky Modes. Journal of Fluid Mechanics, v. 20, pp. 529-554.

Murty, T. S. and Freeman, N. G. (1973) Applications of the concepts of edge waves and numerical modeling to storm surge studies on Lake Huron. Proceedings of 16th Conference. Great Lakes Res. Int. Assoc. Great Lakes Res., Ann Arbor, Mich. pp. 533-548.

Reid R. O. (1958) Effect of Coriolis force on edge waves (I) Investigation of the normal mode. Journal of Marine Research, v. 16, pp. 109 - 144.

Rhines, P. B. (1969a) Slow oscillations in an Ocean of varying Depth. Part I abrupt Topography. Journal of Fluid Mechanics, v. 37, pp. 161-189.

Rhines, P. B. (1969b) Slow Oscillations in an Ocean of Varying Depth. Part II. Islands and Seamounts. Journal of Fluid Mechanics, v. 37, pp. 191-205.

Stokes, G.G. (1846) Report on recent researches in Hydrodynamics. Br. Assoc. Rep. 167p.

Ursell, F. (1951) Trapping Modes in the Theory of Surface Waves. Proceeding of Cambridge Philosophical Society, v. 47, pp. 347-358.

Yanuma, T. and Tsuji, Y. (1998) Observation of edge waves trapped on the continental shelf in the vicinity of Makurazaki Harbor, Kyushu, Japan. Journal of Oceanography, v. 54, pp. 9-18.

GEOMATICS IN TSUNAMI
S.M. Ramasamy, C.J. Kumanan, Brig. R. Sivakumar and Bhoop Singh (eds.), pp. 17-30
New India Publishing Agency, New Delhi (India)
E-mail : spjain_niph@rediffmail.com Web: www.bookfactoryindia.com

3

PHOTOGRAMMETRY APPLICATIONS FOR TSUNAMI DAMAGE ASSESSMENT

M. Dharma Raj

Abstract

An unprecedented tsunami hit the Andaman and Nicobar Islands and the coasts of Andhra Pradesh, Tamil Nadu, Pondicherry and Kerala on the 26th of December 2004. The loss of life and property, though wide spread in all these areas, was very severe in thickly populated low lying coasts, such as Nagapattinam in Tamil Nadu. Survey of India, in association with the Department of Science and Technology, Govt. of India, New Delhi has taken-up the task of carrying out post-tsunami Mapping of the entire east coast of Tamil Nadu using Digital aerial stereo images acquired immediately after the tsunami. The project aims to generate ortho-rectified photographs on 1: 8000 scale, which are to be used for preparing damage assessment maps of the affected coastal areas. This paper discusses various issues related to Control Surveying, Digital Photogrammetry and Large Scale feature extraction that are involved in the mapping process in general and its credibility for tsunami damage assessment in particular.

Keywords: Tsunami, Digital Photogrammetry, Tsunami damage assessment in India.

INTRODUCTION

Photogrammetry

Photogrammetry is defined as the science or art of making reliable measurements by means of photography. Although the term can apply to measurements from ground photographs, modern photogrammetric techniques are most often applied to aerial and satellite images. One of the most common uses of Photogrammetry is the analysis of aerial photographs to extract ground elevations for the production of topographic maps. However, this field also includes the production of digital ortho-photographs and Digital Terrain Models.

Image Acquisition

Photographs

The first step in the production of digital ortho-photographs and terrain models is the capturing of aerial photographs of the land surface. To acquire the images, an aircraft is made to fly over the study area in a straight flight line and photographs are taken in such a way that every ground point appears in at least two successive photographs. The resulting photos are called as stereo pairs and the area of common coverage is called overlap. If the study area of interest requires more than one flight line to have complete coverage, additional flight lines are flown parallel to the original line. Overlap between adjacent flight lines is called sidelap and is used to prevent gaps in coverage. Standard overlap between the adjacent photographs along the flight is 60% and the side lap between two adjacent flight lines is 15 %.

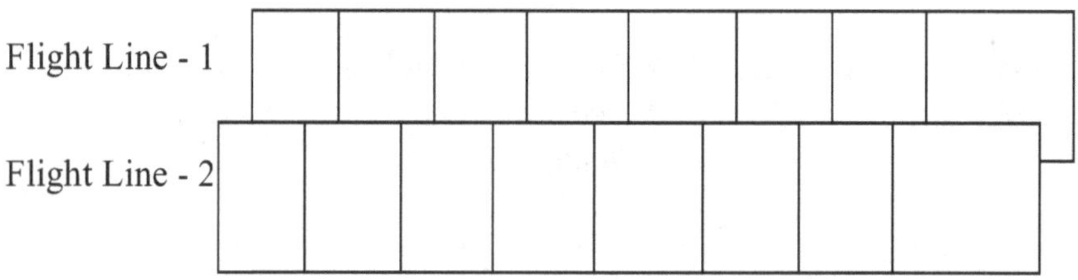

Figure 1: Image Coverage using Overlapping Photographs

Digital Images

Where as a digital Image or a digital photograph consists of a two dimensional matrix with picture elements. Each element is called as pixel. Each pixel is assigned a gray scale value that corresponds to the average intensity of the ground area covered by the pixel. As the resolution of the scanned image decreases, magnification is required to discern the individual pixels with the naked eye.

Digital Photogrammetry

Now a days, most of the photogrammetrists use digital image processing software to extract terrain elevations from aerial photographs. However, the point extraction process is quite similar, with the notable exception that the stereopair is viewed in the computer as a scanned raster image. Using the processing software, the photogrammetrists select and digitize a high density of points. The selected image coordinates, in conjunction with the ground control points, comprise the data points used to construct a Digital Terrain Model. This is referred to as Digital or soft-copy Photogrammetry.

ORIENTATION PROCEDURES

Interior Orientation

Interior orientation defines the internal geometry of a camera or a sensor as it existed at the time of data capture. The variables associated with image space are defined during the process of interior orientation. Interior orientation is primarily used to transform the image pixel coordinate system or other image coordinate measurement system to the image space coordinate system.

Fig.2 illustrates the variables associated with the internal geometry of an image captured from an aerial camera, where *o* represents the principal point and *a* represents an image point.

The internal geometry of a camera is defined by specifying the following variables:

- Principal point
- Focal length
- Fiducial marks and
- Lens distortion

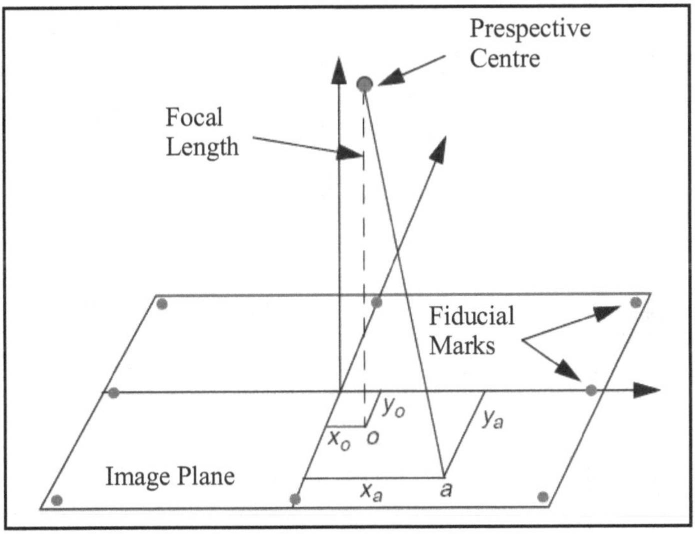

Figure 2: Internal Geometry

Exterior Orientation

Exterior orientation defines the position and the angular orientation associated with an image. The variables defining the position and orientation of an image are referred to as the elements of exterior orientation. The elements of exterior orientation define the characteristics associated with an image at the time of exposure or capture.

The positional elements of exterior orientation include *Xo*, *Yo*, and *Zo*. They define the position of the perspective center (*O*) with respect to the ground space coordinate system (*X*, *Y*, and *Z*). *Zo* is commonly referred to as the height of the camera above sea level, which is commonly defined by a datum.

The angular or rotational elements of exterior orientation describe the relationship between the ground space coordinate system (*X*, *Y*, and *Z*) and the image space coordinate system (*x*, *y*, and *z*). Three rotation angles are commonly used to define angular orientation. They are omega (w), phi (j), and kappa (k). Fig.3 illustrates the elements of exterior orientation.

Omega is the rotation about the photographic x-axis, phi is the rotation about the photographic y-axis, and kappa is the rotation about the photographic z-axis, which are defined as being positive if they are counterclockwise when viewed from the positive end of their respective axis. Different conventions are used to define the order and direction of the three rotation angles. The photographic z axis is equivalent to the optical axis (focal length). The *x'*, *y'*, and *z'* coordinates are parallel to the ground space coordinate system.

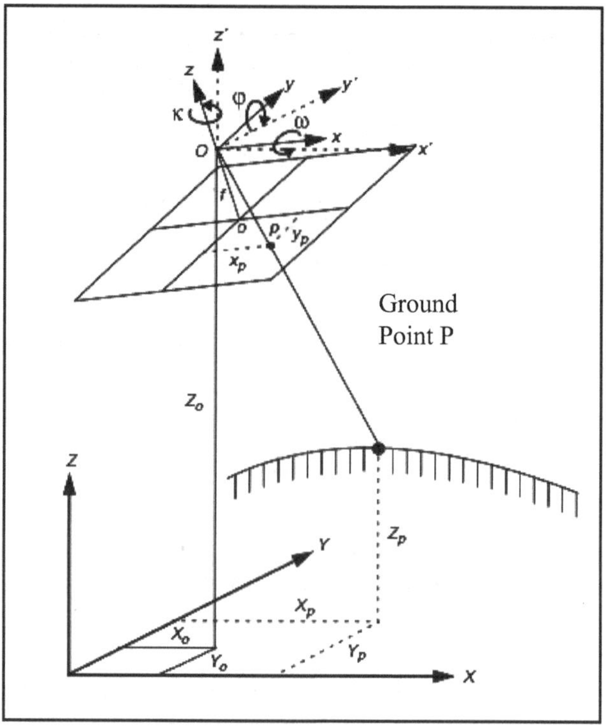

Figure 3: Elements of Exterior Orientation

Using the three rotation angles, the relationship between the image space coordinate system (*x*, *y*, and *z*) and ground space coordinate system (*X*, *Y*, and *Z* or *x'*, *y'*, and *z'*) can be determined. A 3X3 matrix defining the relationship between the two systems is used. This is referred to as the orientation or rotation matrix *M*. The rotation matrix can be defined as follows:

$$M = \begin{bmatrix} m_{11} & m_{12} & m_{33} \\ m_{21} & m_{22} & m_{23} \\ m_{31} & m_{32} & m_{33} \end{bmatrix}$$

The Collinearity Condition Equation

With reference to Fig.3, an image vector '*a*' can be defined as the vector from the exposure station *O* to the image point *p*. A ground space or object space vector *A* can be defined as the vector from the exposure station *O* to the ground point *P*. The image vector and ground vector are collinear, inferring that a line extending from the exposure station to the image point and to the ground is linear.

The image vector and ground vector are only collinear if one is a scalar multiple of the other. Therefore, the following statement can be made:

$$\mathbf{a} = \mathbf{kA}$$

where *k* is a scalar multiple. The image and ground vectors must be within the same coordinate system. Therefore, image vector "a" comprises the following components:

$$a = \begin{bmatrix} x_p - x_o \\ y_p - y_o \\ -f \end{bmatrix}$$

where *xo* and *yo* represent the image coordinates of the principal point.

Similarly, the ground vector can be formulated as follows:

$$A = \begin{bmatrix} X_p - X_o \\ Y_p - Y_o \\ Z_p - Z_o \end{bmatrix}$$

In order to have the image and ground vectors to be within the same coordinate system, the ground vector must be multiplied by the rotation matrix *M*. The following equation can be formulated:

$$\mathbf{a = k\,MA}$$

where

$$\begin{pmatrix} x_p - x_o \\ y_p - y_o \\ -f \end{pmatrix} = kM \begin{pmatrix} X_p - X_o \\ Y_p - Y_o \\ Z_p - Z_o \end{pmatrix}$$

The above equation defines the relationship between the perspective center of the camera/sensor exposure station and ground point *P* appearing on an image with an image point location of *p*. This equation forms the basis of the collinearity condition that is used in most of the photogrammetric operations. The collinearity condition specifies that the exposure station, ground point and its corresponding image point location must all lie along a straight line, thereby being collinear. Two equations comprise the collinearity condition.

$$\left[\frac{m_{11}(X_p - X_{o1}) + m_{12}(Y_p - Y_{o1}) + m_{13}(Z_p - Z_{o1})}{m_{31}(X_p - X_{o1}) + m_{32}(Y_p - Y_{o1}) + m_{33}(Z_p - Z_{o1})}\right]$$

$$\left[\frac{m_{21}(X_p - X_{o1}) + m_{22}(Y_p - Y_{o1}) + m_{23}(Z_p - Z_{o1})}{m_{31}(X_p - X_{o1}) + m_{32}(Y_p - Y_{o1}) + m_{33}(Z_p - Z_{o1})}\right]$$

One set of equations can be formulated for each ground point appearing on an image. The collinearity condition is commonly used to define the relationship between the camera/sensor, the image and the ground.

PHOTOGRAMMETRIC SOLUTIONS

Digital photogrammetry is used for many applications, ranging from ortho-rectification, automated elevation extraction, stereopair creation, feature collection, highly accurate point determination and control point extension.

For any of the afore mentioned tasks to be undertaken, a relationship between the camera/sensor, the image(s) in a project and the ground must be defined. The following variables are used to define the relationship:

- Exterior orientation parameters for each image
- Interior orientation parameters for each image
- Accurate representation of the ground

Depending on the input data provided, photogrammetric techniques such as space resection, space forward intersection and bundle block adjustment are used to define the variables required to perform ortho-rectification, automated DEM extraction, stereopair creation, highly accurate point determination and control point extension.

Bundle Block Adjustment

A block of images contained in a project is simultaneously processed in one solution. There are several models for block triangulation. Bundle block adjustment uses the collinearity condition as the basis for formulating the relationship between image space and ground space.

A bundled solution is computed including the exterior orientation parameters of each image in a block and the X, Y, and Z coordinates of tie points and the adjusted Ground Control Points (GCP). A statistical technique known as least squares adjustment is used to estimate the bundled solution for the entire block while also minimizing and distributing error.

In order to understand the concepts associated with bundle block adjustment, an example comprising two images with three GCPs whose X, Y, and Z coordinates are known is used. Additionally, six tie points are available. Fig.4 illustrates the photogrammetric configuration.

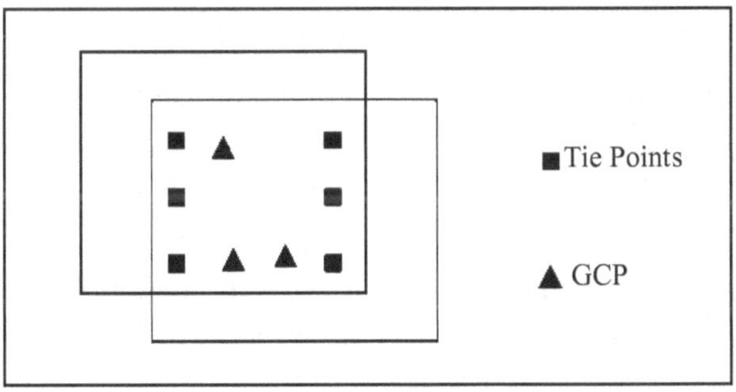

Figure 4: Photogrammetric Configuration

Forming the Collinearity Condition Equations

For each measured GCP, there are two corresponding image coordinates (x and y). Thus, two collinearity equations can be formulated to represent the relationship between the ground point and the corresponding image measurements. In the context of bundle block adjustment, these equations are known as observation equations.

If a GCP has been measured on the overlapping areas of two images, four equations can be written: two for image measurements made on the left image of the pair and two for the image measurements made on the right image of the pair. Thus, GCP *A* measured on the overlap areas of left image and right image has four collinearity formulae.

$$x_{a1} - x_o = -f \left[\frac{m_{11}(X_A - X_{o1}) + m_{12}(Y_A - Y_{o1}) + m_{13}(Z_A - Z_{o1})}{m_{31}(X_A - X_{o1}) + m_{32}(Y_A - Y_{o1}) + m_{33}(Z_A - Z_{o1})} \right]$$

$$y_{a1} - y_o = -f \left[\frac{m_{21}(X_A - X_{o1}) + m_{22}(Y_A - Y_{o1}) + m_{23}(Z_A - Z_{o1})}{m_{31}(X_A - X_{o1}) + m_{32}(Y_A - Y_{o1}) + m_{33}(Z_A - Z_{o1})} \right]$$

$$x_{a1} - x_o = -f \left[\frac{m'_{11}(X_A - X_{o1}) + m'_{12}(Y_A - Y_{o1}) + m'_{13}(Z_A - Z_{o1})}{m'_{31}(X_A - X_{o1}) + m'_{32}(Y_A - Y_{o1}) + m'_{33}(Z_A - Z_{o1})} \right]$$

$$y_{a1} - y_o = -f \left[\frac{m'_{21}(X_A - X_{o1}) + m'_{22}(Y_A - Y_{o1}) + m'_{23}(Z_A - Z_{o1})}{m'_{31}(X_A - X_{o1}) + m'_{32}(Y_A - Y_{o1}) + m'_{33}(Z_A - Z_{o1})} \right]$$

One image measurement of GCP *A* on Image 1: x_{a1}, y_{a1}

One image measurement of GCP *A* on Image 2: x_{a2} , y_{a2}

Positional elements of exterior orientation on Image 1: X_{o1} , Y_{o1} , Z_{o1}

Positional elements of exterior orientation on Image 2: X_{o2} , Y_{o2} , Z_{o2}

If three GCPs have been measured on the overlap areas of two images, twelve equations can be formulated, which includes four equations for each GCP.

Additionally, if six tie points have been measured on the overlap areas of the two images, twenty-four equations can be formulated, which includes four for each tie point.

The previous scenario has the following unknowns:

- Six exterior orientation parameters for the left image (i.e., X, Y, Z, omega, phi, kappa)
- Six exterior orientation parameters for the right image (i.e., X, Y, Z, omega, phi and kappa) and
- X, Y and Z coordinates of the tie points. Thus, for six tie points, this includes eighteen unknowns (six tie points times three X, Y, Z coordinates).

The total number of unknowns are 30. The overall quality of a bundle block adjustment is largely a function of the quality and redundancy in the input data. In this scenario, the redundancy in the project can be computed by subtracting the number of unknowns, 30, by the number of knowns, 36. The resulting redundancy is six. This term is commonly referred to as the degrees of freedom in a solution.

Once each observation equation is formulated, the collinearity condition can be solved using an approach referred to as least squares adjustment.

Least Squares Adjustment

Least squares adjustment is a statistical technique that is used to estimate the unknown parameters associated with a solution while also minimizing error within the solution. With respect to block triangulation, least squares adjustment techniques are used to:

- Estimate or adjust the values associated with exterior orientation
- Estimate the X, Y, and Z coordinates associated with tie points
- Estimate or adjust the values associated with interior orientation
- Minimize and distribute data error through the network of observations

Data error is attributed to the inaccuracy associated with the input GCP coordinates, measured tie point and GCP image positions, camera information and systematic errors.

The least squares approach requires iterative processing until a solution is attained. A solution is obtained when the residuals associated with the input data are minimized.

The least squares approach involves determining the corrections to the unknown parameters based on the criteria of minimizing input measurement residuals. The residuals are derived from the difference between the measured (i.e., user input) and computed value for any particular measurement in a project. In the block triangulation process, a functional model can be formed based upon the collinearity equations

The functional model refers to the specification of an equation that can be used to relate measurements to parameters. In the context of photogrammetry, measurements include the image locations of GCPs and GCP coordinates, while the exterior orientations of all the images are important parameters estimated by the block triangulation process.

The residuals, which are minimized, include the image coordinates of the GCPs and tie points along with the known ground coordinates of the GCPs. A simplified version of the least squares condition can be broken down into a formula as follows:

$$V = AX - L, \text{ including a weight matrix } P$$

Where

V = the matrix containing the image coordinate residuals

A = the matrix containing the partial derivatives with respect to the unknown parameters, including exterior orientation, interior orientation, XYZ tie point, and GCP coordinates

X = the matrix containing the corrections to the unknown parameters

L = the matrix containing the input observations (i.e., image coordinates and GCP coordinates)

The X matrix contains the corrections to the unknown exterior orientation parameters. The X matrix is calculated in the following manner:

$$X = (A^t PA)^{-1} A^t PL$$

Where

X = the matrix containing the corrections to the unknown parameters t

A = the matrix containing the partial derivatives with respect to the unknown parameters

t = the matrix transposed

P = the matrix containing the weights of the observations

L = the matrix containing the observations

The difference between the initial measurements and the new estimates is obtained to provide the residuals. Residuals provide preliminary indications of the accuracy of a solution. The residual values indicate the degree to which a particular observation (input) fits with the functional model. For example, the image residuals have the capability of reflecting GCP collection in the field. After each successive iteration process, the residuals become smaller until they are satisfactorily minimized.

The results from the block triangulation are then used as the primary input for the following tasks:

- Stereo pair creation
- Feature collection
- Highly accurate point determination
- DEM extraction
- Ortho-rectification

GCPs

The instrumental component of establishing an accurate relationship between the images in a project, the camera/sensor and the ground are GCPs. GCPs are identifiable features located on the Earth's surface whose ground coordinates in X, Y, and Z are known. A full GCP has associated with its X,Y, and Z (elevation of the point) coordinates. Horizontal control only specifies the X, Y, whereas vertical control only specifies the Z. The following features on the Earth's surface are commonly used as GCPs:

- Intersection of roads
- Utility infrastructure (e.g., fire hydrants and manhole covers)
- Intersection of agricultural plots of land
- Survey benchmarks

DIGITAL TERRAIN MODEL

A triangular irregular network (TIN) model is typically used to digitally represent the terrain surface (Fig.5). A TIN is a triangulated mesh constructed on the (x,y) locations of a set of data points. To form the TIN, a perimeter around the data points is first established, called the convex hull. To connect the interior points, triangles are created with all internal angles as nearly equiangular as possible. This procedure is called Delaunay triangulation. By including the dimension of height (z) for each triangle vertex, the triangles can be raised and tilted to form a plane. The collection of all such triangular planes forms a representation of the land surface terrain in a considerable degree of detail.

Additional elevation data such as spot elevations at summits and depressions and break lines can also be included in the TIN model. Break lines represent significant terrain features like a lake or cliff that cause a change in slope; TIN triangles do not cross break lines.

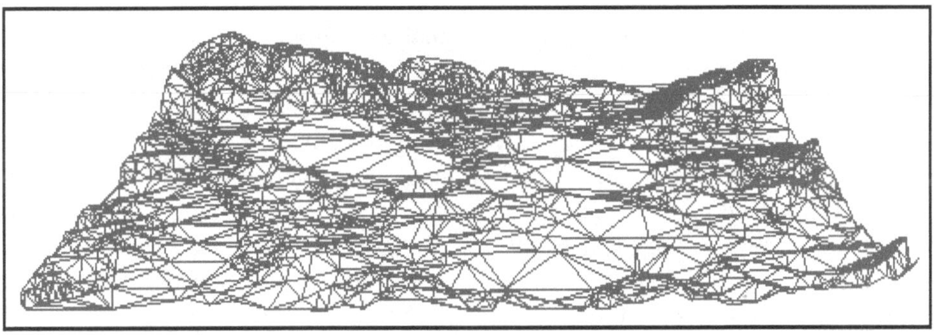

Figure 5: Triangular Irregular Network Surface Representation

The elevation points extracted from the stereomodel can also be used to create a raster (grid) model of terrain elevations. A land surface representation in the raster domain is called a digital elevation model (DEM).

ORTHO-RECTIFICATION

Ortho-rectification is the process of removing geometric errors inherent within photography and imagery (Fig.6). The variables contributing to geometric errors include, but are not limited to:

- Camera and sensor orientation
- Systematic error associated with the camera or sensor
- Topographic relief displacement
- Earth curvature, etc.

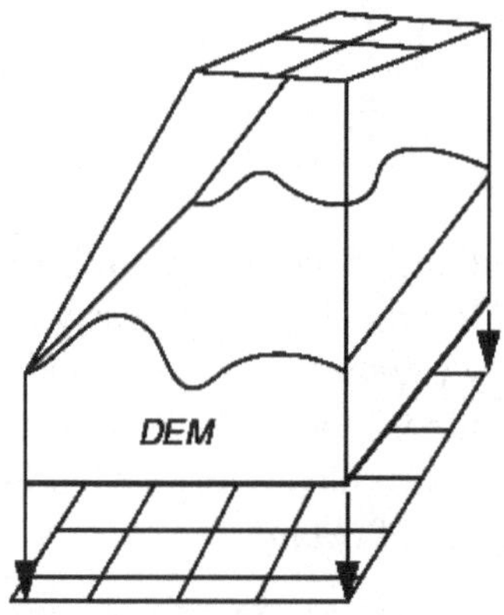

Ortho-rectified image

Figure 6: Ortho-rectification

By performing block triangulation or single frame resection, the parameters associated with camera and sensor orientation are defined. Utilizing least squares adjustment techniques, during block triangulation, minimize the errors associated with camera or sensor instability. Additionally, the use of self-calibrating bundle adjustment (SCBA) techniques along with Additional Parameter (AP) modeling accounts for the systematic errors associated with camera interior geometry. The effects of the Earth's curvature are significant if a large photo block or satellite imagery is involved. They are accounted for during the block triangulation procedure by setting the proper option. The effects of topographic relief displacement are accounted for by utilizing a DEM during the ortho-rectification procedure. The ortho-rectification process takes the raw digital imagery and applies a DEM and triangulation results to create an ortho-rectified image. Once an ortho-rectified image is created, each pixel within the image possesses geometric fidelity. Thus, measurements taken off an ortho-rectified image represent the corresponding measurements as if they were taken on the Earth's surface (Fig.6).

An image or photograph with an orthographic projection is one for which every point looks as if an observer was looking straight down at it, along a line of sight that is orthogonal (perpendicular) to the Earth. The resulting ortho-rectified image is known as a digital ortho-image (Fig.7).

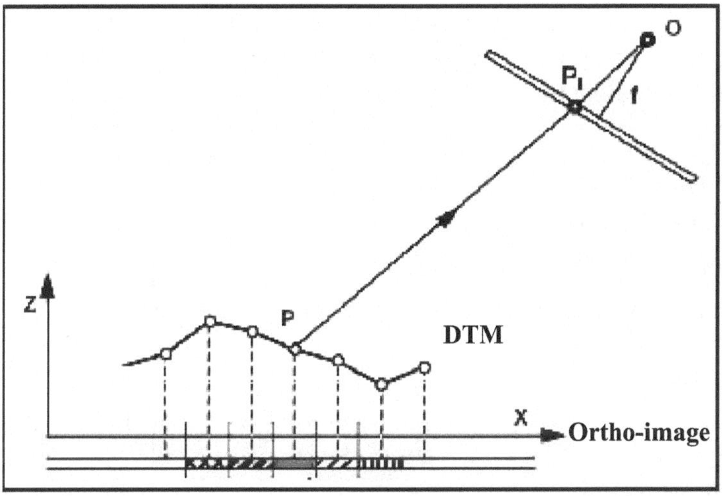

Figure 7: Digital Ortho-image - Finding Gray Values

Relief displacement is corrected by taking each pixel of a DEM and finding the equivalent position in the satellite or aerial image. A brightness value is determined for this location based on resampling of the surrounding pixels. The brightness value, elevation and exterior orientation information are used to calculate the equivalent location in the ortho-image file.

In Fig.7

P = ground point

P_1 = image point

O = perspective center (origin)

X, Z = ground coordinates (in DTM file)

f = focal length

In contrast to conventional rectification techniques, ortho-rectification relies on the digital elevation data, unless the terrain is flat. Various sources of elevation data exist, such as a DEM automatically created from stereo image pairs.

METHODOLOGY FOR POST-TSUNAMI MAPPING

The Tamil Nadu and Pondicherry coast line of about 900 km length was captured by Digital aerial Photography, immediately after the tsunami. The width covered in the photography is about 2 km across the coast. These Digital images of about 1700 photographs were made available to SOI for post-tsunami Mapping. The area along the coast of about 1800 Sq. Km was divided into several blocks as per coast line alignment for the purpose of providing Ground control. Each block was separately controlled for block adjustment. Planimetric GCPs were provided using Dual frequency GPS receivers by rapid static method with 1 base receiver and multiple rovers. Height control was provided using Digital Levels. The observations were processed for determination of Co-ordinates on WGS-84 and UTM systems. Digital Photogrammetry software was used to further process the images for

- Block adjustment
- Model Formation
- Automatic DEM Generation
- Ortho-rectification and generation of ortho-photos on scale 1:8000

CONCLUSION

According to the concept note published by the National Disaster Management Division of Ministry of Home Affairs, Govt. of India, (NDM 2005), GIS mapping and land use planning are essential for proper disaster planning, Emergency Management, Rescue and Response, Insurance Needs and Community Education. The present Mapping exercise is an effort towards fulfilling this national goal. The various agencies that are involved in R & D for assessment, prevention, mitigation and early warning of tsunamis will utilize the base maps/ortho-photos, generated through this project.

REFERENCES

Karl Kraus (1993) Photogrammetry. Ferd. Dummler, Verlag, Bonn.

Eric Tate (1998) Photogrammetry Applications in Digital Terrain Modeling and floodplain mapping. Published in the Internet.

National Disaster Management Division (2005) Government of India, Ministry of Home Affairs, March 2005, Prevention/Protection and Mitigation from Risk of tsunami Disasters - A Concept Note, published in the Internet.

Leica Photogrammetry suite (2003) User Manual, Ver. 8.7.

TSUNAMI –
DISASTER ASSESSMENT

GEOMATICS IN TSUNAMI
S.M. Ramasamy, C.J. Kumanan, Brig. R. Sivakumar and Bhoop Singh (eds.), pp. 31-36
New India Publishing Agency, New Delhi (India)
E-mail : spjain_niph@rediffmail.com Web: www.bookfactoryindia.com

4

INDIAN TSUNAMI (2004): INITIATIVES BY NRDMS, DEPARTMENT OF SCIENCE AND TECHNOLOGY, INDIA

Brig. R. Siva Kumar and Bhoop Singh

Abstract

The major subduction of the Indian plate and the related high magnitude earthquake (>9M) occurred in Sumatra region, triggered a major tsunami on 26, Dec 2004 which caused massive disasters in many coastal regions of South Asian countries including India, killing tens of thousands of life and causing property loss of several million rupees. Duly realizing the credentials of Geospatial / Geomatics technology and the expertise of NRDMS, an all India coordinated programme was conceived by NRDMS on tsunami disaster assessment and got it executed through six specialized institutions working on geomatics technology in the country. In the programme, the tsunami disaster in terms of natural, physical and human resources were assessed using geomatics technology and village wise digital databases were generated which will form a potential baseline information in coastal area.

Keywords: Indian Ocean Tsunami 2004, DST Initiatives, Disaster Assessment, Mitigation.

INTRODUCTION

On December 26, 2004, the great tsunami ravaged many coastal regions of South Asia, including the entire East Coast of India and Andaman - Nicobar Islands killing thousands of lives and damaging properties worth of several hundred crores. The tsunamis are generated predominantly by large sub-marine earthquakes. This particular tsunami was trigged by the subduction of Indian plate under Burmese micro plate along a North- South fracture of 1200 km length, 100 km breadth and a subduction of 15 meter depth. This subduction has caused a massive earthquake of very high magnitude (>9M) leading to the displacement of huge quantum of ocean water and as a result of which giant waves were generated and traveled both eastwards and westwards. Such tsunami waves traveled with the wave length of 600 – 700 km, amplitude of 30 cm to 1.5 meter and a speed of 700 –800 km / hr in the ocean gained a huge height of 8 to 19 meters along the coast and caused devastation along Tamil Nadu, Andhra Pradesh and southern parts of Kerala coast, besides Andaman and Nicobar islands

This major tsunami has raised a number of the administrative and scientific issues. Also, it has alerted the concerned State Governments and the entire scientific community to evolve comprehensive mitigative action plan for combating such disasters. This paper, deals about the R&D efforts initiated by DST just after the tsunami.

INITIATIVES BY DST

Just after the 26th December 2004 tsunami, a multi-disciplinary team of experts was sent to visit the tsunami affected areas of Tamil Nadu coast for assessing the extent of damage due to the disastrous tsunami. On the basis of the report of the team, it was decided to undertake detailed studies on the pattern of damage and prepare damage maps in GIS environment, which can be efficiently used for relief and rehabilitation operations for tsunami and can also be used for futuristic coastal issues in general and tsunami management in particular.

OBJECTIVES

The following objectives were set for the said all India coordinated tsunami assessment programme in six affected areas in Andaman and the east coast of India:

- ➢ Preparation of damage assessment maps on natural, physical and human resources in GIS environment on 1:8,000 scale including
  - Determination of soil salinity and land use / land cover changes, coastal erosion and deposition along with beach profiles changes in the tsunami affected areas.
  - Assessment of contamination in portable ground water resources and preparation of sustainable plan and also
- ➢ Preparation of risk maps for future planning strategies.

INSTITUTIONS AND THE STUDY AREA (Fig.1)

In order to achieve the set objectives of the programme, DST initiated six specific R&D projects on pilot scale (Fig.1). The name of the institutions and their study area are given in the following table:

S.No	Principal Investigators / Institutions	Study area
1.	Dr. P. Chakrabarti, West Bengal, State Council of Science & Technology, Kolkata	Parts of Andaman & Nicobar Islands.
2.	Dr. K. Mruthyunjaya Reddy, Andhra Pradesh State Remote Sensing Application Center, Department of Planning, Government of Andhra Pradesh, Hyderabad.	Parts of Prakasam/ Nellore coast, Andhra Pradesh
3.	Dr. S. Kaliappan, Institute of Remote Sensing, Anna University, Chennai.	Chennai coast & Surroundings, Tamil Nadu
4.	Dr. G. Victor Rajamanickam, SASTRA, Thanjavur.	Cuddalore coast, Tamil Nadu
5.	Dr. SM. Ramasamy, Center for Remote Sensing, School of Geosciences, Bharthidasan University, Trichy	Nagapattinam coast, Tamil Nadu
6.	Dr. N. Chandrasekar, Center for Marine Science and Technology, MS University, Tirunelveli	Kanyakumari coast, Tamil Nadu

METHODOLOGY

Just after tsunami, NRSA has flown digital aerial photography on the approximate scale of 1:8,000. The IRS, 1C/1D PAN merged data of pre-tsunami period was used for comparison purposes. DST has coordinated the procurement of data products from Survey of India for smooth handling by the all concerned institutions. The orthophotos of the affected areas were generated through Survey of India and subsequently provided to all the groups for analysis and demarcation of the extent of damage and large scale mapping. The main aim was to collect all the above data pertaining to various disasters and create GIS data base to derive solutions for various management plans related to tsunami.

MONITORING MECHANISM

In February, 2005, DST sanctioned all the six projects to the concerned institutions / principal investigators for undertaking detailed investigations. The satellite data and aerial photographs were made available to them for preparing large scale maps and other damage assessment of the selected areas. In order to monitor the progress of all the concerned groups from time to time, regular meetings were conducted by DST with the concerned groups and few experts to meet the defined objectives of the entire project.

During the entire exercise, close interactions were made with the NRSA to identify the satellite data requirement (pre and post-tsunami) and its smooth flow to all the PIs. In addition to this, a standardized data collection matrix for natural resources as well as matrix for infrastructure damage was prepared, which was followed by all the PIs while undertaking detailed studies in their respective study areas. At present, all the projects are in their final stage of completion. DST in close association of all the PIs has brought out a special volume on tsunamis entitled (Geomatics in Tsunami –the present book) highlighting the research work done by all the participating agencies with DST's support, as well as inviting articles from other experts on some specific themes.

1. State Council of Science & Technology, Kolkata. 2. Andhra Pradesh State Remote Sensing Application Center, Government of Andhra Pradesh. 3. Institute of Remote Sensing, Anna University, Chennai. 4. Shanmuga Arts Science Technology and Research Academy, Thanjavur. 5. Center for Remote Sensing, School of Geosciences, Bharthidasan University, Trichy. 6. Center for Marine Science & Technology, MS University, Tirunelveli.

Figure 1: Areas studied by different institutions for Tsunami disaster assessment under NRDMS sponsored All India Coordinated program

COASTAL RESOURCES MANAGEMENT AND CONSERVATION

In addition, as a part of NRDMS Programme, concerted efforts are being made to understand the practices of coastal resources management and their conservation. More than dozen pilot projects have been already supported by NRDMS to various academic institutions to understand the following aspects:

- Assessment of renewable and non-renewable resource potential.
- Assessment of coastal stability – measures for stabilization.
- Assessment of tourism potential, carrying capacity of the environment and plans for infrastructure development.
- Assessment of energy generation potential – wave and wind energy.
- Assessment of vulnerability of the coast to – cyclones, floods, typhoons etc. and contingency planning.
- Assessment of impact of human activity on the coast and control measures.

TRANSFER OF TECHNOLOGY AND CAPACITY BUILDING

Capacity building is the process by which individuals, groups, organizations strengthen their ability to carry out their functions and achieve desired results over a time. Under NRDMS Programme, several geo-spatial tools and techniques for data management and analysis have been developed in the areas of natural resources management and natural disaster assessment (tsunamis and landslides). Development of training kits, modules for different target groups in map making and handling spatial databases and also information retrieval to local planning have been the main targets under this programme. The research work done under tsunami project, would also form a strong information base to facilitate planners for taking relevant decisions to minimize the impact of coastal disasters in general and that of tsunami in particular.

INTERNATIONAL COOPERATION

As a part of ongoing Indo-Norwegian Project on "Human Resource Development in applied research for assessment, prevention, mitigation and early warning of tsunamis and landslides in India", concerted efforts are being made to upgrade the skill and expertise of our Indian Scientists on par with the International standards in the area of geohazards (tsunamis and landslides) through a technical collaboration with International Center for Geo-hazards, Oslo, Norway. Tsunami modeling – wave propagation model are being developed using Indian data, so that capability of our scientists is built up to further carry on the research in this area. Large scale mapping and risk assessment is another area where capabilities are being developed under this programme. The scientists of Indian institutions, who are participating in the programme would be trained on the advanced technologies related to the management of geo-hazards (tsunamis and landslides).

ISSUES AND SCOPE OF FURTHER WORK

The great tsunami of December, 2004 has compelled to think over and assess the capabilities of the administration as well as scientific community to handle the situation in the event of such crisis. In this context the following basic issues are to be dealt in detail:

- Prediction and fore warning of tsunamis
- Tsunami wave propagation in the ocean and its modifications along the coasts.
- Tsunami vulnerabilities of Indian coasts.
- The impacts of tsunami on natural resources and infrastructure.
- Societal issues – Human behavior and response at the time of event.

GEOMATICS IN TSUNAMI
S.M. Ramasamy, C.J. Kumanan, Brig. R. Sivakumar and Bhoop Singh (eds.), pp. 37-62
New India Publishing Agency, New Delhi (India)
E-mail : spjain_niph@rediffmail.com Web: www.bookfactoryindia.com

5

S & T INPUT: EARTHQUAKE AND TSUNAMI EFFECTS, ANDAMAN ISLANDS, INDIA

Parthasarathi Chakrabarti, Sanjoy Nag, Subrata B. Dutta, Susmita Dasgupta and Niladri Gupta

Abstract

A very strong shallow focus submarine earthquake (of magnitude 9.0) occurred off the west coast of Sumatra Island due to the movement of Indian Plate and Burma Plate on 26th December 2004 at 06:18:53 IST. A vast area around its epicentre was rocked including Andaman and Nicobar Islands and parts of the eastern and western coasts of the Indian mainland. This earthquake had generated a global tsunami, which had also pounced on the Andaman and Nicobar Islands and the coasts of Andhra Pradesh, Tamil Nadu and Kerala and caused severe damage to the coastal population and infrastructure. Considering the devastation of this tsunamigenic earthquake, the Department of Science & Technology, Govt. of India has launched a coordinated programme for imparting S & T inputs with futuristic approach to support rehabilitation programme in the affected areas on a sustainable manner. The main objective of the programme is to generate geospatial databases on damages and risk zonation using pre and post-tsunami remote sensing (RS) data along with collateral attribute data in GIS platform. This paper speaks on the damage assessment (both natural / physical resources and human resources) caused by the earthquake and tsunami covering parts of Andaman Islands along with strategies for risk management and constraints and alternatives for micro level planning.

Keywords: Tsunamigenic earthquake, Damage Assessment, Remote Sensing and GIS, Andaman - Nicobar.

INTRODUCTION

The major earthquake (of magnitude 9.0) occurred on the Sunda belt on 26th December 2004 at 6:18:53 IST, hit the sea off the west coast of Northern Sumatra.

The epicentre of this earthquake was at 3.24° N: 95.82° E and the focus was at 37 Km below the sea floor.

This earthquake is the total effect of three events as described by E. R. Musson, British Geological Survey, "the initial slip that occurred first to the west of Sumatra's northern tip and two subsequent slips to its north triggered by the first. The total force released was enough to jolt the entire planet". Subsequently, studies made by USGS and Geological Survey of India (GSI) regarding the epicentres of 26-12-2004 and 28-03-2005 and their aftershocks reveal that they were confined to the west of the 96° E Fracture zone (Table-1, Fig.1) on the roof of the down going Indian Ocean slab / Indian Plate suggesting a break in tectonic connection between the segments on either side of this fracture zone and further readjustment (Banerjee et al 2005).

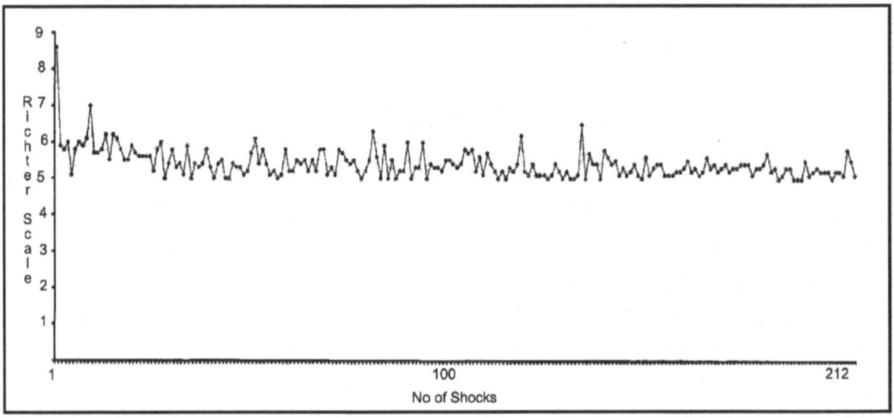

Figure 1: Frequency and magnitude of 212 tremors felt in and around Andaman from 26th Dec. 2004 to 5th Feb. 2005 (Source: IMD, DST, Andaman)

It may be mentioned that the 26-12-2004 event resulted in widespread adjustment in the Andaman - Nicobar Islands (Fig.2a, b). Survey of India indicates that the northern part of the Andaman – Nicobar Islands has uplifted with a eastward tilt e.g. in Diglipur, North Andaman (uplift between 0.5 to 0.8 m), whereas in Port Blair, South Andaman (south east displacement by1.15 m approx.) and subsidence of southern most tip of Great Nicobar (Indira Point) by 4.25 m MSL (New Zealand Herald, 02.03.2005).

The other event associated with this earthquake was the most devastating 'tsunami' in history. The deadly tsunami waves caused widespread damage in eleven countries bordering the Indian Ocean with propagation of waves in E – W direction as the submarine fracture was caused by the earthquake in a near N – S direction.

"Zooming in" to the Indian coast the most severe attack of tsunami waves was faced by the Andaman – Nicobar Islands due to their proximity to the epicentre.

Table 1: Details of shocks of Magnitude 5.0 and above on Richter Scale West of 96°E Fracture Zone (Source: IMD, DST, Andaman)

Sl No	Date	Time (IST)	Latt	Long	Magni-tude	Approx. Geographic Location
1	26/12/2004	0629	03.34	94.13	8.6	West Coast of Sumatra
2		0719	05.39	94.42	5.9	NW of Sumatra
3		0746	12.38	92.51	5.8	West of M. Andaman
4		0752	08.84	92.53	6.0	50 Km SSW of C. Nicobar
5		0805	04.00	94.18	5.1	West Coast of Sumatra
6		0806	12014	93.01	5.8	25 Km SE of Baratang
7		0822	12051	92.59	6.0	West of M. Andaman
8		0829	03.08	94.20	5.9	West Coast of Sumatra
9		0839	13.81	92.97	6.1	North of Diglipur
10		0951	06.98	92.81	7.0	120 Km West of C. Bay
11		1152	10.62	92.32	5.7	30 Km South of Hutbay
12		1237	10.30	93.76	5.7	100 Km SSE of Hutbay
13		1308	13.12	93.05	5.8	Close to SE of N. Andaman
14		1450	06.10	90.95	6.2	60 Km West of Indira Point
15		1548	08.95	93.73	5.5	55 SSE of Car Nicobar
16		1549	12.65	94.69	6.2	90 Km NE of Barren Island
17		1635	14.80	92.42	6.1	100 Km NNW of N.Andaman
18		1740	13.81	92.25	5.8	70 Km NW of N. Andaman
19		1742	11.60	92.45	5.5	20 Km SW of Port Blair
20		1927	02.63	94.60	5.5	530 Km S of Indira Point
21		2019	13.64	92.44	5.9	40 Km W of N. Andaman
22		2037	03.70	94.02	5.7	350 Km S of Indira Point
23		2042	06.70	93.02	5.6	90 Km S of Indira Point
24	27/12/2004	1616	13.63	93.10	5.4	NE of N. Andaman
25		2017	13.49	91.46	5.9	150 Km W of N. Andaman
26	28/12/2004	0044	11.57	91.54	5.3	90 Km W of Port Blair
27	29/12/2004	0744	12.17	91.35	5.4	100 Km W of Port Blair
28		1159	12.94	92.16	5.1	120 Km NNW of Port Blair
29		1201	13.21	92.15	5.2	90 Km W of N. Andaman
30	30/12/2004	2328	12.62	92.10	5.5	70 Km S of M. Andaman
31	31/12/2004	1124	11.54	91.93	5.8	60 Km SW of Port Blair
32	08/01/2005	1802	11.90	92.70	5.3	30 Km N of Port Blair
33	12/01/2005	0426	12.40	91.00	5.1	West of M. Andaman
34	14/01/2005	2152	09.30	93.30	5.0	North Coast of N. Andaman
35		2318	12.20	92.50	5.1	80 Km NW of Port Blair
36	15/01/2005	1317	15.30	92.20	5.4	NW of Port Blair
37	27/01/2005	2054	07.60	96.60	5.1	Andaman Sea
38	31/01/2005	0310	11.50	91.70	5.1	180 Km W of Port Blair
39	02/06/2005	0924	14.00	93.40	5.6	N Coast of N. Andaman

The southeastern and southwestern coasts of mainland India were also affected by the catastrophic tsunami covering parts of Andhra Pradesh, Tamil Nadu and Kerala.

Considering the devastation to the lives and properties due to earthquake and tsunami, the NRDMS Division, Department of Science & Technology, Government of India has launched a coordinated programme for imparting S&T inputs with futuristic approach to facilitate conceptualization and operationalisation of

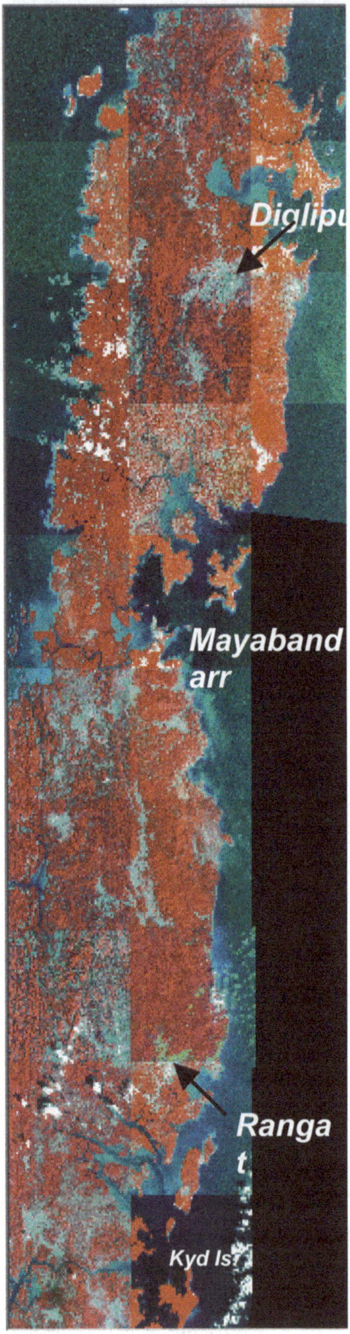

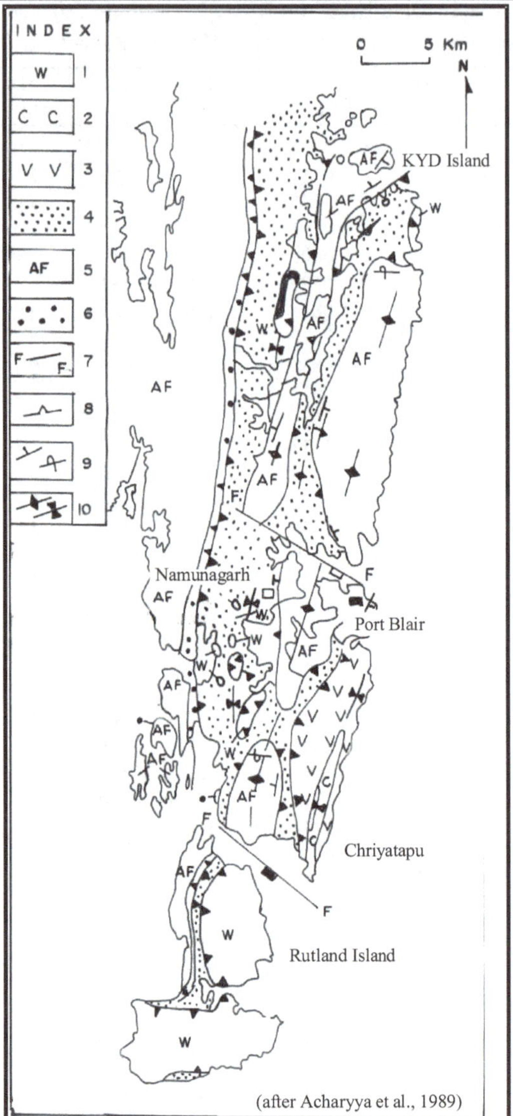

1. Undifferentiated ultramafic and mafic rocks
2. Cumulates 3. Mafic Volcanics
4. Melange (ophiolite derived clastics, ophiolitic rocks and olistromal sediments)
5. Andaman Flysch 6. Archipelago Gp 7. Fault
8. Thrust 9. Strike and dip
10. Axial trace of antiform / window, Synform Klippe.

Figure 2a: North and Middle Andaman as viewed by IRS P6 MX & LISS III Sensors

Figure 2b: Geological Map – parts of Andaman Islands.

alternative options and associated limitations of land use in micro level planning with regional perspective.

Under this programme six test sites were included covering parts of Andhra Pradesh, Tamilnadu and Andaman islands with the following objectives:

- Preparation of damage assessment maps (on appropriate scale) on natural / physical resources and human resources using remote sensing data and GIS technique.
- Study on soil salinity and land use / land cover changes including coastal erosion and deposition for the tsunami affected areas.
- Assessment of contamination in potable ground water resources and preparation of sustainable plan.
- Generation of risk maps for future planning strategies.

METHODOLOGY

Towards achieving the above aim and objectives of the programme, the following methodology was adopted:

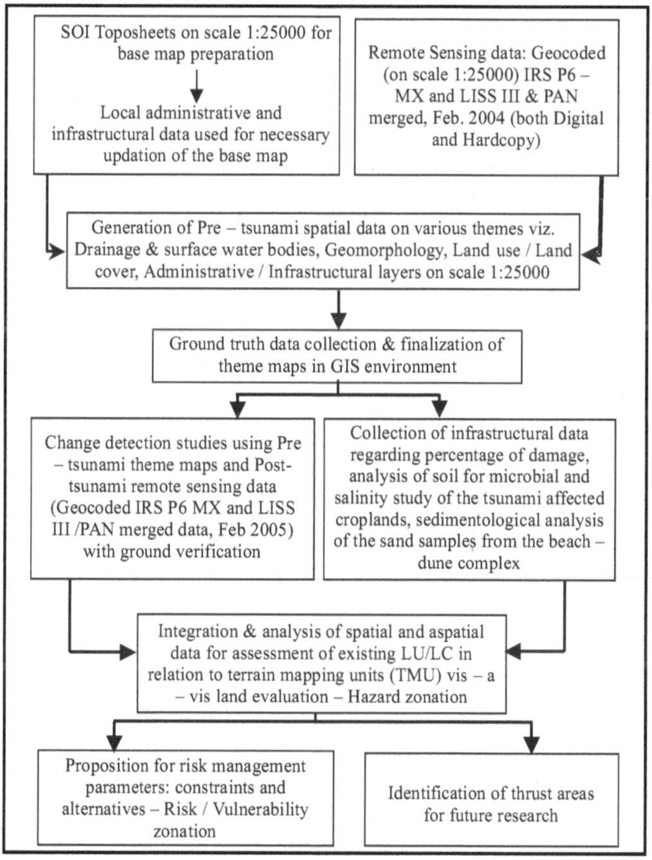

STUDY AREA

The Andaman island is located in Indian Ocean within 92^0 30' & 93^0 30' E Longitudes and 11^0 & 14^0 N Latitudes. More specifically, the study area includes the North and Middle Andaman. The highest point of the study area is the

Table 2: Administrative setup, Andaman District

District	Sub Division	Tehsil
Andaman	Mayabandar	Diglipur
		Mayabandar
		Rangat
	South Andaman	Ferrargunj
		Port Blair
		Little Andaman*
* Not yet given the status of tehsil		

Table 3: Panchayats in Mayabandar Subdivision

Tehsil		
Diglipur	**Mayabandar**	**Rangat**
Gram Panchayat		
Radhanagar #	Mayabandar*	Shivapuram #
Laxmipur	Pokadera*	Nimbutala*
Madhupur	Rampur*	Rangat #
Sitanagar	Pahalgaon	Parnasala #
Subhashgram	Chainpur #	Dasarathpur #
Diglipur	Basantipur	Sabari*
Ramkrishnanagar	Harinagar	Bakultala #
Keralapuram #	Swadeshnagar	Urmilapur
Shibpur #		Kaushalyanagar #
Kishorinagar		Kadamtala
Ramnagar*		Uttara #
Nabagram #		Long Island*
Kalighat #		Sundargarh #
		Nilambur #
* indicates settlements along coast # indicates settlements along major creeks		

Saddle Peak (▲737 m) SW of Kalipur, North Andaman. The study area covers three Tehsils of Andaman district viz. Rangat, Mayabandar and Diglipur under Mayabandar sub – division (Table–2 & 3). The Tehsil / Sub - division headquarters are well connected with Port Blair, the Capital of the Union Territory, through Andaman Trunk Road (ATR) with two ferry crossings at Nilambur and Uttara and a bridge over Austin Strait connecting Mayabandar and Diglipur. However, the bridge was damaged during the 26th December 2004 earthquake. In addition to the ATR, almost all the Panchayats and other villages are linked with either black topped or metalled road as well as by waterways. Local harbours are present at Aerial Bay/ Diglipur, Mayabandar and Rangat.

Figure 3: Pillow Lava and Agglomerate, Mayabandar, Middle Andaman

Figure 4: Rocky Coast (Sheet Rock, Sandstone), Durgapur, North Andaman

Geology

The Andaman & Nicobar islands are essentially composed of Late Jurassic to Early Eocene rocks (viz. ophiolites and sedimentary rocks of diverse lithology) which have been deformed by numerous deep seated faults and thrusts and also have been intruded by ultramafic igneous rocks (Fig.2a, 2b). The Andaman ophiolite (ANO) constitutes the upper structural unit of dismembered ophiolite suite, predominantly made up of igneous rocks like pillow basalt, agglomerate and other volcanic rocks (Fig.3). The sedimentary melange (Fig.4) tectonically flooring these ophiolite mélange consists of argillaceous and olistostromal beds (algal limestone) locally associated with rhythmites and minor sandstone [Lipa Formation (Acharyya 1997) = Baratang Formation (Chatterjee 1967, Ray 1982, Roy 1983)].

Structurally, the western fringe of the Andaman island represents a thrust contact trending N – S (Fig.2b). The axial traces of the antiform and synform are almost parallel to the trend of the western thrust contact (Acharyya 1997).

Figure 5a: Mud Volcano, Nayagarh, Baratang Island

Figure 5b: Pseudo columnar structure in ejected material with sedimentary rock fragments

Figure 5c: Volcanic cone with fresh flow

It is worth mentioning here that there are eleven numbers of mud volcanoes in the north and middle Andaman Islands along the crestal part of the main geanticline (Poddar 1954, Banerjee 1975). Out of these eleven mud volcanoes six are present in the Baratang Island, South Andaman, two in Middle Andaman and three in North Andaman. The mud volcanoes are invariably present where the country rocks are of the Baratang Formation (Chatterjee 1967). The Baratang Formation consists mainly of dark gray shale with pyrite and minor sandstone and algal limestone.

Two major mud volcanoes are located near the Jarwa creek around Rajatgarh and Nayagarh villages along the Prithi fault (NW – SE) and Jarwa creek fault (Banerjee 1975 and Basu et al 2005). The ejected matters with columnar structures consist of gray to grayish blue mud along with rock fragments from the underlying older sedimentary complex (Fig.5a, b, c).

Geomorphology

Geomorphologically, the study area includes various terrain units (Meijerink, 1988) viz.

1. Structural – Denudational hills and slopes
2. Hill wash deposits / Glacis
3. Valleys
4. Coastal landform features:
   a) cliffed coast (e.g. Mayabandar)
   b) rocky coast (Fig.3, 4, 6)
   c) sandy beach (Fig.7)
   d) intertidal mud flat with / without mangroves (Fig.8).
   e)Fluvio – tidal flat (e.g. Dasarathpur, Rangat)

It may be mentioned here that main emphasis is given to classify the coastal tract of the study area in human orientation as per the objectives of the programme considering "Man as a Geomorphic agent" (Verstappen 1987). In this respect, the existing land use / land cover pattern plays an important role. We revert to this aspect later.

Land Use

The Andaman and Nicobar Islands are well known for its forest resources (about 92% of the geographical area) ranging from mangroves to dense evergreen forest (Fig.9). Almost all along the coastal stretch and creeks rich mangrove swamp zone exists (Fig.8). In response to the tropical hot and humid climate with abundant rainfall, all the major islands support very luxuriant vegetation.

Figure 6: Rocky Coast, Amkunj, Middle Andaman

Figure 7: Sandy Beach, Kalipur, North Andaman

Figure 8: Inter tidal mud flat with / without Mangrove, Mayabandar, Middle Andaman

Figure 9: Hill with Evergreen Forest and Mangrove swamp zone, Nilambur, Middle Andaman

As per the Forest Survey of India (1999), out of the total area under forest cover, dense forests with a crown density of 40% constitute 85.6%, open forests with crown density of less than 40%, constitute 1.7% and mangrove constitute 12.7%. District wise and category wise forest cover is furnished in Table-4a, b.

Table-4a: Details of Forest cover in Andaman & Nicobar Islands (Area in sq km)

District	Dense	Open	Mangrove	Total
Andaman	4864	90	929	5883
Nicobar	1651	35	37	1723
Total	6515	125	966	7606

Table-4b: Details of Recorded Forest in Andaman & Nicobar Islands (Area in sq km)

District	Reserved Forests	Protected Forests	Total Forests
Andaman	2928.76	2699.86	5628.62
Nicobar	-	1542.07	1542.07
Total	2928.76	4241.93	7170.69

Forestry in these islands dates back to 1883, at that time there was no regular working plan and harvesting of timber was confined to a few species only *viz.,* Padauk, Silvergrey and Koko.

Scientific management of forest started with the preparation of first working plan by F.H. Todd in 1906, followed by working plans prepared by Banington in 1914 and H.S. Dean in 1935, but due to various reasons, prescriptions of these working plans could not be adhered to. The working schemes prepared by Chengappa between1936 to 1939 were organized according to the prescriptions. After Second World War and partition of India, Chengappa prepared working plan in 1952 but due to various socio-political reasons, the prescriptions of his working plans could not be implemented. Subsequent working plan prepared by Shri J. C. Verma in

1960, was also not implemented. The current management practice involves a silvicultural system based on irregular shelter wood system with floating periodic blocks. Regeneration technique as evolved recently is called "The Andaman Canopy Lifting shelterwood system". The system provides for gradual lifting of canopy in order to allow adequate light filtered through the canopy. Forest plantation, (mainly Teak and Pema) is in vogue at places along with coconut plantation in agriculture farms. Considering the importance of mangroves to protect the land from sea / storm surges, mangrove plantation also exists (Fig.10a, b).

Figure 10a: Mangrove Swamp with plantation during Low Tide

Figure 10b: Mangrove Swamp with plantation during High Tide

Figure 11: Hill Slope with Terrace Cultivation, Rampur, North Andaman

Figure 12: Sandy Beach with Berm face, Ramnagar, North Andaman – A Beach Resort

In the seaward periphery of mangrove zone, coral reef is present (16 Families, 31 genera and 82 species – Ramakrishna 1996)

Sensu - stricto, with rehabilitation of the refugees from, the then, East Pakistan (now Bangladesh) after independence, the land use pattern of the islands, started changing and presently one may find croplands in the hill slopes and valleys (Fig.11) or along the coastal tract (e.g. beach resort, eco - park etc.) wherever settlements are present (Fig.12). However, agriculture is mainly rain fed and main crop is paddy. In addition, vegetables are grown by the dwellers specifically in the North Andaman.

DISCUSSION ON DAMAGE ASSESSMENT

Comparative study of pre and post-tsunami remote sensing data supported with ground truthing, interview of the local people and ancillary data collection of the study area demand discussion on damage assessment caused by earthquake and tsunami in two different sectors viz. physical/ natural environment and human orientation.

The effects of the severe earthquake are very much exemplified on ground as well as on man made structures in the form of cracks, slips/ fractures, collapsed structure, etc. It is reported by the GSI (2004) that the intensity of earthquake remained same in North, Middle and South Andaman. However, the devastation in South Andaman (especially in Port Blair and Little Andaman) is more due to the combined effects of earthquake and tsunami.

North and Middle Andaman

In Mayabandar Subdivision, North and Middle Andaman (refer Table-2 & 3) the damage due to earthquake is more conspicuous than the tsunami wave attack.

The main reasons lie with the physiographic configuration of the area and very specific location of the settlements in the higher ground away from the HWL (High water line). However, damage due to earthquake includes toppling of household articles (*viz.* T.V., refrigerator, furniture, etc.) in the area. Some major findings are furnished below in Table-5.

Table 5: Earthquake and Tsunami Damages.

Tehsil	Gram Panchayat	Findings/ Remarks
Diglipur	Diglipur	At Aerial Bay due to upliftment, shifting of HWL away from the land by about 2.0 m; no damage in the Jetty; tsunami waves touched the road, about 3.0 m from the present water level.

Contd...

Table 5 Contd...

	Kalipur	Very gently sloping sandy beach with flattened dunes at places; tsunami waves toppled the dune; 'Keya' trees present all along the dune/ upper beach face absorbs the thrust of waves; no infrastructural damage due to absence of settlements / inhabitated area along the beach dune complex. Earthquake damage is more prominent; cracks developed in the newly built double storied Panchayat building at Kalipur (Fig.13).
	Ramnagar	The R.C. bridge on Ramnagar creek damaged due to earthquake; new iron bridge under construction about 100m downstream of the old R. C. bridge (Fig.14) Sandy beach with 'berm' face (Fig.12); covered by 'Keya' trees and other species; as such no damage identified due to tsunami waves in this beach resort; no effect on settlements located about 1.0 km from the coastline.
	Mohanpur	Deep cracks of about 2.0 m (E – W) in the blacktopped road (Fig.15) and damage of some weak tin roofed huts.
	Austin Strait	Major damage in the Austin bridge –Translational movement in NE – SW direction (cf. Billings, 1968) present - middle span shifted towards NE by 0.45 m and adjacent span gone down by 0.10m (Fig.16).
	Mayabandar	No such major damage due to earthquake reported. Tsunami wave effect restricted with raising of sea water level and flooding mainly in the tidal flats.
	Pokadera (Baludera)	Tsunami wave effect very much restricted within 200 – 250 m. from the HWL; Sea water ingressed in the cultivable land, Agricultural Farm (coconut plantation) (Fig.17)
Mayabandar	Rampur	Earthquake damage – a newly built up two storied building collapsed due to failure of load bearing RC columns of the ground floor. The 2nd floor is now resting on the ground; may be an example of liquefaction - area of future research on engineering and geological aspects. (Fig.18).
	Karmatang	Rocky / bouldery beach with some sandy patches: sandy beach steep in nature with 'berm' and low height dune with 'Ipomia biloba', 'Keya' trees / mangroves; tsunami wave toppled the dunes at places but restricted within 25 m (Fig.19); insignificant or no damage in the 'Eco – park'.

Contd...

Table 5 Contd...

Rangat	Shivapuram & Nimbutala	The seaward buffer zone of the ATR (Andaman Trunk Road) affected at places due to tsunami wave, though rocky coast - trees uprooted and soil / road materials washed away (Fig.20).
	Dasarathpur	Tsunami wave attack predominant; the reclaimed portion of the intertidal flat being waterlogged (Fig.21, 22a,b) – cultivable land and a portion of the settlement (3 / 4 huts) damaged (Fig.23) - an example of unscientific anthropogenic activities.
	Rangat	Minor repairable damage in the Jetty and adjacent spur due to earthquake as well as tsunami wave; however, total collapse of a 3 storied RCC building - may be due to faulty construction.
		Tsunami wave entered into the fishers colony located within 50 m from HWL - minor infrastructural damage; however, 50 / 60 fishing boats damaged as reported.
	Kadamtala	Minor cracks in the floor of the houses; however, damage of household articles (viz. TV, refrigerator, almirah etc.) toppled during shocks at 6.30 hrs and 10.00 hrs as reported by the people (the IMD data reveals that at 9.51hrs there was another major shock of magnitude 7.0 - refer Table-1)
	Sundargarh	Step like ground subsidence with cracks (E – W) on the black topped road near Nayagarh.
	Nilambur	Tsunami water flashed into the cultivable land through Jarwa creek near the mud volcano at Nayagarh; local villagers reported that immediately after the earthquake two mud volcanoes started erupting; however no loss of life and property as no settlement nearby; eruptions of mud took place as a single large explosion at 6.35 hrs at Jarwa creek and at 11.00 hrs at Rajatgarh on 26.12.2004. The concomitant gas flow at the Jarwa creek mud volcano caught fire on 27.12.2004 producing orange / red coloured flames to a height of 60 to 90 cm and finally died down on 31.12.2004 (Basu et al 2005)
		Though rise of water level in Middle Strait flooded the road, however, no damage occurred due to presence of broad mangrove swamp zone.

Figure 13: Earthquake damage to double storied Panchayat building, Kalipur, North Andaman

Figure 14: Construction of Iron Bridge 100 m downstream of damaged R..C. bridge, Ramnagar, North Andaman

Figure 15: Deep cracks (E-W) on black topped road, Mohanpur, North Andaman

Figure 16: Translational movement (NE-SW) in middle span of Austin Bridge, Mayabandar, Middle Andaman

Figure 17: Tsunami affected coast, Pokadera (Baludera), Mayabandar Sub Division, Middle Andaman

Figure 18: Damaged two storied building; failure of load bearing RC column(inset), Rampur, Middle Andaman

Figure 19: Insignificant damage by tsunami, Karmatang, Middle Andaman

Figure 20: Erosion in the seaward buffer of the ATR by tsunami, Chitrakut, Middle Andaman

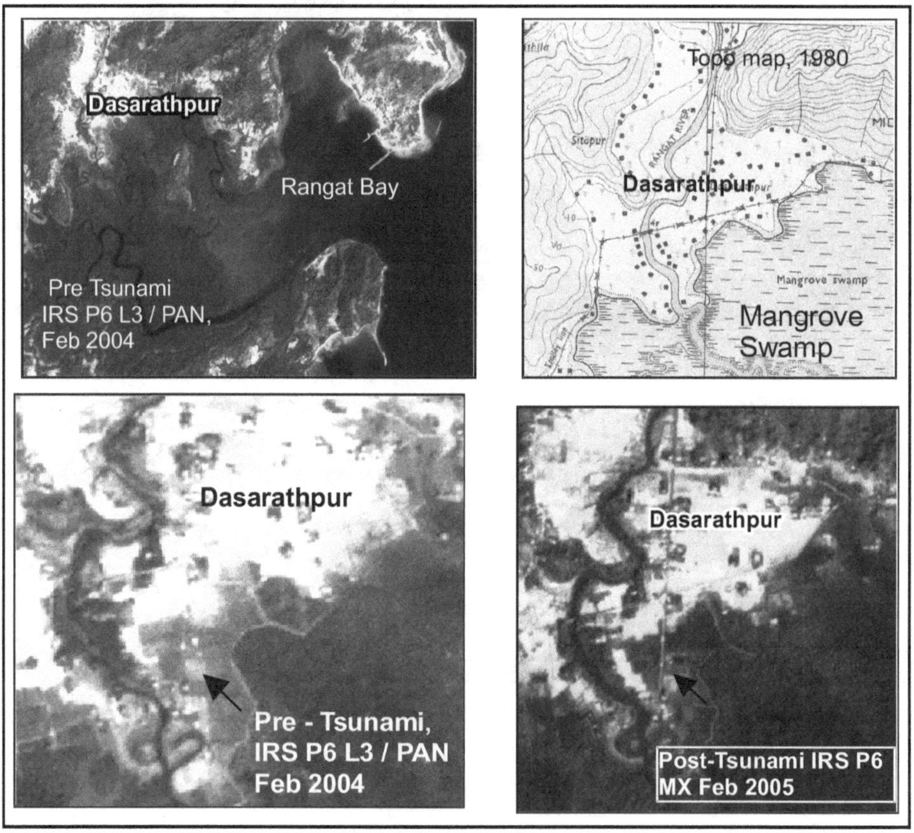

Figure 21: Pre and Post-tsunami Land use / Land cover changes as viewed by IRS P6 - comparable to topo map

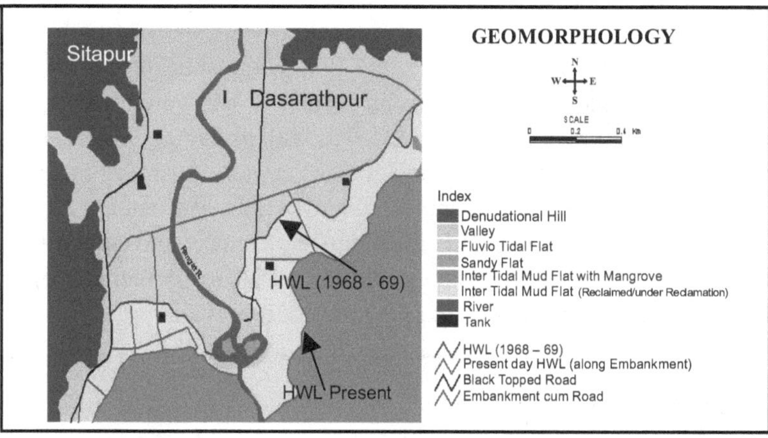

Figure 22a: Geomorphological map around Dasarathpur, Rangat

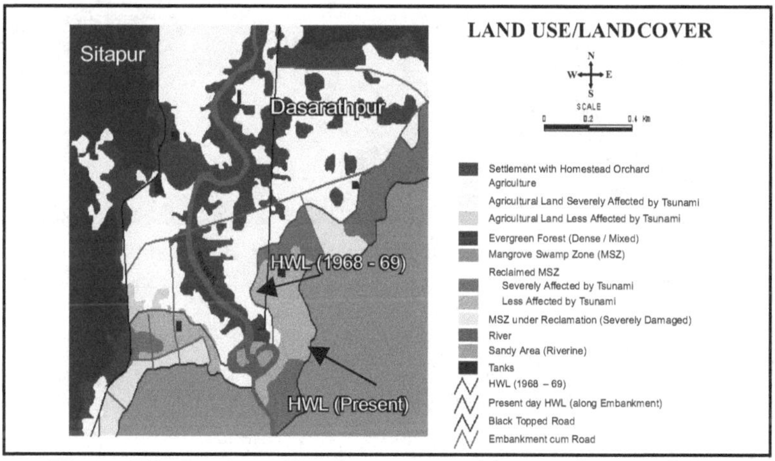

Figure 22b: Land use / Land cover map with Hazard Susceptibility around Dasarathpur Rangat

Figure 23: Damaged cropland and coconut plantation due to tsunami Dasarathpur Rangat

South Andaman

In South Andaman subdivision, in the Port Blair and Ferrarganj tehsils the effect of tsunami is more conspicuous in terms of sea water ingression in the croplands and agricultural farms (Fig.24, 25) as well as damage of the ATR around Ograbaraij and Sipighat area. The new road is under construction in the same alignment with an increase in height by about 1m, as the aerial extent of the intertidal flat increased due to rise of the HWL caused by earthquake and tsunami. It may be mentioned that the settlements in the low-lying areas (mostly reclaimed) are worst affected by the tsunami and water logging condition prevails.

Figure 24: Tsunami affected waterlogged area, Sipighat, Port Blair, South Andaman

Figure 25: Post-tsunami regular inundation of agri farm during high tide, Ograbaraij – Sipighat, Port Blair, South Andaman

Analysis of Soil Samples

Physico - chemical and microbial analysis of the soil samples collected from the tsunami affected and non – affected portions of the cultivable lands of the study area (Table-6a) reveal that

- pH and electrical conductivity (EC) of the tsunami affected soil have increased due to accumulation of soluble salt from sea water. Organic carbon has decreased in the tsunami affected soil due to the accumulation of sand / silt.
- Microbial Biomass Carbon (MBC) content in the tsunami affected soil ranged between 37 – 122 µg/g and it decreased in the tsunami affected soil, may be due to presence of excess amount of salt (as reflected by EC value).

Table-6a: Physico–chemical and microbial study of tsunami affected and non - affected soils, North & Middle Andaman

Soil No.	Location	Soil Condition	pH	EC (ds/m)	OC (g/kg)	MBC (µg /g)	MBC OC (%)
1	Pokadera, Mayabandar	Affected Not affected	7.5 6.9	11.2 1.17	3.5 3.5	47 79	1.34 2.26
2	Baludera, Mayabandar	Affected Not affected	8.4 7.9	13.5 2.18	8.7 8.2	122 187	1.40 2.28
3	Rampur, Mayabandar	Affected Not affected	6.9 6.2	14.6 2.25	3.5 4.1	51 111	1.46 2.71
4	Karmatang, Mayabandar	Affected Not affected	6.8 6.4	15.4 1.51	3.2 4.4	53 131	1.66 2.98
5	Dasarathpur, Rangat	Affected Not affected	6.2 5.9	19.9 2.46	2.8 4.1	41 127	1.46 3.10
6	Nayagarh, Baratang	Affected Not affected	6.5 6.3	19.5 1.94	2.9 4.9	37 141	1.28 2.88
7	Ograbaraij, Port Blair	Affected Not affected	6.3 6.1	22.8 2.14	6.6 7.5	104 164	1.58 2.19
8	Chouldari, Port Blair	Affected Not affected	6.4 5.8	23.8 2.07	5.4 6.6	73 178	1.35 2.70

The absolute amount of biomass–C at one time is not indicative of whether soil organic matter is increasing or decreasing and should be compared with a related soil parameter, such as soil organic-C (OC) to provide a measure of organic matter dynamics as well as soil disturbances. In the tsunami affected soil MBC/OC (%) has decreased and ranged between 1.28 – 1.66.

In totality, the sea water ingression deteriorated the soil quality parameters like microbial biomass-C due to accumulation of sand and presence of excess amount of salt. Analysis of exchangeable metal content of the soil samples (Table-6b) reveals that Na, K, Ca and Mg have increased in the tsunami affected soil.

Table-6b: Variation in exchangeable metals in the tsunami affected Andaman-affected soils

Soil No.	Location	Soil condition	mg/kg Na	mg/kg K	mg/kg Ca	mg/kg Mg
1	Pokadera, Mayabandar	Affected Not affected	2654.2 104.0	249.6 104.0	3600.0 453.6	770.4 45.0
2	Baludera, Mayabandar	Affected Not affected	900.0 170.2	134.0 96.0	1948.0 312.0	386.4 129.6
3	Rampur, Mayabandar	Affected Not affected	2578.3 104.0	276.9 76.0	3416.0 375.8	935.0 100.0
4	Karmatang, Mayabandar	Affected Not affected	2348.3 98.0	92.0 62.4	3024.0 312.6	799.2 165.0

Sedimentological Analysis of Sand Samples from Beach Dune Complex

The granulometric analysis of the sand samples (Folk and Ward 1957) collected from the present day beach face and dunes (toppled by the waves during tsunami) along the eastern coast of North and Middle Andaman Islands are furnished in Table-7. In general, the sand samples are medium to coarse grained, moderately well sorted with fragments of shells (turritella, gastropods, pelycipods), corals etc.

Table 7: Granulometric analysis of the sand samples

Location	Nature of Samples	Grain size parameters				
		Modal Class	Mean	Sorting	Skewness	Kurtosis
Kalipur (Lower Beach Face)	Fine to Medium Calcareous Sand	Unimodal	Medium Sand	Well Sorted	Coarse Skewed	Mesokurtic
Kalipur (Upper Beach Face)	Fine to Medium Calcareous Sand	Unimodal	Medium Sand	Moderately Well Sorted	Coarse Skewed	Leptokurtic
Baludera (Beach)	Medium to Coarse Calcareous Sand	Trimodal	Coarse Sand	Moderately Sorted	Symmetrica l	Very Platykurtic
Baludera (Tsunami deposited sand)	Medium Calcareous Sand	Unimodal	Medium Sand	Moderately Well Sorted	Symmetrica l	Mesokurtic
Karmatang (Beach)	Medium Calcareous Sand	Unimodal	Medium Sand	Moderately Well Sorted	Symmetrica l	Platykurtic
Karmatang (Dune sand, tsunami affected)	Medium Calcareous Sand	Unimodal	Medium Sand	Moderately Well Sorted	Coarse Skewed	Mesokurtic
Ramnagar (Beach)	Medium Calcareous Sand	Unimodal	Medium Sand	Moderately Sorted	Coarse Skewed	Platykurtic
Ramnagar (Dune sand, tsunami affected)	Medium Calcareou s Sand	Unimodal	Medium Sand	Moderately Sorted	Coarse Skewed	Mesokurtic

The unimodal nature of the samples, in general, indicates unidirectional sediment transport both during normal tide / wave action as well as tsunami period. This also corroborates the fact that both the regular tides / waves and tsunami are from the same direction i.e. from the east of the study area.

Potable Groundwater Resources

In Andaman Islands springs are ubiquitous due to physiographic configuration. In general, spring water is being collected in the settling tanks and is supplied through pipeline to the settlements (Fig.11). So, there is no such possibility of contamination of sea water into potable water due to tsunami.

It may be mentioned that due to availability of spring water, use of dug wells are very limited. Local villagers of Kadamtala gram panchayat, Rangat tehsil reported that the dug wells with water became dry and dry dug wells became full of water after the earthquake. Similarly, the local villagers of Mohanpur gram panchayat, Diglipur tehsil reported that dug wells became dry during earthquake and rejuvenated afterwards. Such phenomenon are quite possible in the dug wells located in the valley flat or in the hill slopes as hydraulic continuity appears to have been disturbed due to shaking of the granular zone.

FUTURE OUTLOOK

In the context of the devastating effects of the tsunamigenic earthquake in the Andaman – Nicobar Islands and coasts of mainland, India, the following recommendations (both long term and short term) are made with futuristic approach considering the interdependency of various physical parameters and anthropogenic influences.

Scientific Aspects

In the context of readjustment of the tilting effect due to tectonic movement -

- Detailed geological, geophysical and applied geomorphological studies may be carried out to identify the zone of weakness (e.g. fault scarps, fissures), tilted surface (viz. uplifted or downthrown) and accretional or erosional beach faces for generation of hazard susceptibility / vulnerability maps with micro zonation (Chakrabarti and Nag 2002). Beach survey programme in specific areas should be carried out at regular intervals for generation of prototype data to develop a data bank on the 'base case scenario' in terms of 'form – process – material' interaction in the beach dune complex zone for setting up models for erosional and accretional zones through space and time(Chakrabarti 1991).
- Regular monitoring of the ejections of the mud volcanoes may reflect on the tectonic activities in the Andaman Islands.

- Engineering geological studies may be carried out to safeguard the infrastructural developments e.g. bridges, roads, buildings etc.
- Change detection studies of the terrain features and land use / land cover practices based on Geo-environmental Mapping concept for tackling the 'site specific' problems (Chakrabarti 2004).
- Comparative analysis of the flow pattern of the springs (especially those which are in use for water supply to the settlements) before and after the earthquake should be carried out.

Applied Aspects

- Strict imposition of the Coastal Regulation Zone (CRZ) Notification, Ministry of Environment and Forests, Govt. of India (as amended upto 25.01.2005) to avoid / reduce casualties, loss of property etc. due to storm surges, super cyclones or tsunami.
- Protection of the existing mangrove forests as well as intensive mangrove plantation in the mud flat zone to be encouraged (especially bordering the settlements), instead of reclamation of mangrove swamp zone. The clayey substrate supports mangrove colonization and ultimately that will act as buffer zone to withstand the thrust of storm / cyclone waves as well as to nourish the intertidal zone as the stilt roots of the mangroves act as silt catcher.
- Public education / awareness programme to be conducted at regular intervals for the coastal population (especially to the fishers colony) to familiarise them regarding protection of mangroves, disadvantages of aggravated reclamation of mangrove swamp zone, CRZ notification, seismic proof and tsunami safe constructions etc.

CONCLUSIONS

The submarine earthquake originated due to subduction of the Indian plate has generated the tsunami on 26th December 2004. Andaman being nearer to the epicentre of the earthquake is affected both by the earthquake tremors and tsunami. Due to this plate movement the North Andaman was uplifted and South Andaman has gone down as reported. Reflection of this phenomenon is exemplified in the North Andaman by damage due to earthquake where as combined effect of earthquake and tsunami is present in the South Andaman.

It is worth mentioning here that the reclaimed intertidal mangrove swamp zone is mostly affected in the South Andaman reflecting the unscientific approach towards encroachment of nature's own built buffer zone for such a devastation. This may be considered as the best example of "Technology Push" rather than

"Application Pull" which leads to impart / implement technology not appropriate to the terrain condition and social settings → "Ethno –science" to be improved by "Techno-science" (i.e. Technology Push) for more responsive land use / land cover practices by protecting the nature's own built system towards minimisation of adverse effects of natural hazards in relation to changing socio economic environment.

In this regard, rational and optimal utilization of remote sensing data for periodic updation of the thematic aspects of terrain units and subsequent integration and analysis of the spatio – thematic information with the collateral attribute data (to be collected by *insitu* / conventional methods) in GIS environment may be considered as "Application Pull" to specify the natural and man induced domains towards societal benefits and eco-system preservation (i.e. sustainable land use planning) through

- Model solution for tackling the natural hazards.
- Identification of thrust areas for future S & T inputs.

ACKNOWLEDGEMENT

The authors express their gratitude to the Department of Science & Technology, Govt. of India for entrusting the work to this department with fund support. The authors would like to place on record the encouragement and support provided by the Working Chairman, West Bengal State Council of Science & Technology and the Principal Secretary, Department of Science & Technology, Govt. of West Bengal for accomplishing the work in a time bound manner. Sincere thanks to Dr. K. Chakrabarti, Reader, Department of Agricultural Chemistry and Soil Science and Sri. Sudipta Lahiri, Director, Marine Division, Geological Survey of India for providing necessary support to carry out soil and sedimentological analyses. The contribution of Dr. P. Bhattacharya, Jr. Scientist, WBSCS&T for soil analysis is duly acknowledged. The authors convey their thanks to Sri Rishikesh and Sri D. Bhadra, Scientists, Deptt. of Science & Technology, Andaman for their co-operation during ground truthing and data collection.

REFERENCES

Acharyya, S.K. (1997) Stratigraphy and tectonic history reconstruction of the Indo-Burma-Andaman Mobile Belt. Indian Journal of Geology, v. 69(3), pp. 211-234.

Banerjee, A. (1975) Mud volcanoes of the Andaman Islands. Indian Journal of Earth Sciences, v. 2(1), pp. 11-17.

Banerjee, P.K. et al (2005) An alternative tectonic interpretation of crustal configuration below the Andaman Sea. Newsletter, Marine Wing, Geological Survey of India, v. XVIII (2) & XIX (1), pp. 6-11.

Basu, P. C. et al (2005) 26th December, 2004 Earthquake of Sumatra and mud volcanic eruption in the Baratang Island, Middle Andaman. Newsletter, Marine Wing, Geological Survey of India, v. XVIII (2) & XIX (1), pp. 32-34.

Billings, M.P. (1968) Structural Geology, Prentice Hall, New Delhi, pp. 125-129.

Chakrabarti, P. (1991) Process-response-system analysis in the Macrotidal estuarine and Mesotidal coastal plain of eastern India. Memoirs of Geological Society of India, No.22, pp. 165-181.

Chakrabarti, P. (2004) Geoinformatics for naturel ressources management vis - a - vis environmental justice. ISPRS, XX Congress, Istanbul, v. B VII, Pt 7/ 3, pp. 427-432.

Chakrabarti P. and Nag S. (2002) Geoinformatics: Spatio-thematic information for natural hazard – disaster mitigation. IAPRS & SIS, NRSA, Hyderabad, India v. 34, Pt. 7, pp. 1198-1201.

Chatterjee, P.K. (1967) Geology of the main islands of the Andaman area. Proceeding of Symposium on Upper Mantle Project, NGRI, Hyderabad, India, pp. 348-360.

Folk, R. L. and Ward, W. C. (1957) Brazos river bar; a study in the significance of grain size parameters. Journal of Sedimentary Petrology, v. 27, pp. 3-26.

Meijerink, A.M.J. (1988) Data acquisition and data capture through Terrain Mapping Units, ITC Journal Special Issue. No. 1, pp. 23-44.

Poddar, M.C. (1954) Mud volcanoes of South Baratang Island (Andaman Group). Indian Minerals, v.8, pp.251-256.

Ramakrishna (1996) Coastal Living Resources. Integrated Coastal Zone Management – A Manual, Dept. of Environment, Govt. of West Bengal. pp. 131-139.

Ray, K.K. (1982) A review of the geology of Andaman & Nicobar Islands. Geological Survey of India Miscellaneous Publication, No. 41, pp. 110-125.

Roy, T.K. (1983) Geology and hydrocarbon prospects of Andaman – Nicobar basin, In.: Petroliferous basins of India, Petroleum Asia Journal, v. 1(6), pp. 37-50.

Verstappen, H. Th. (1987). Geomorphology, resources, environment and the developing world; International Geomorphology, Part-I. John Wiley, pp. 45-62.

GEOMATICS IN TSUNAMI
S.M. Ramasamy, C.J. Kumanan, Brig. R. Sivakumar and Bhoop Singh (eds.), pp. 63-76
New India Publishing Agency, New Delhi (India)
E-mail : spjain_niph@rediffmail.com Web: www.bookfactoryindia.com

6

DAMAGE ASSESSMENT AND MAPPING OF TSUNAMI AFFECTED AREAS IN PARTS OF NELLORE COAST, INDIA

K. Mruthyunjaya Reddy, A.Nageswara Rao,
V. Raghu and A. V. Subba Rao

Abstract

A large tsunami triggered due to a massive earthquake of 9.3M in the offshore of Sumatra on 26th December 2004 created havoc in several countries of the Indian Ocean. The deadly tsunami waves lashed the towns adjoining the coastline of several countries including Indonesia, Thailand, Malaysia, India and Sri Lanka causing more than 1,50,000 deaths. The earthquake occurred along the plate boundary is marked by a subduction zone between the Indian plate and the Burmese micro plate. In Andhra Pradesh, six coastal Districts were affected and amongst them four south coastal districts experienced maximum damage. In the present study, Tuppalapalem (vagarru revenue village) of Vakadu mandal, Nellore District was selected as a case study for discussion in the article. The objective of the study was to make an inventory of the damages to the natural and physical resources, by preparing pre and post-tsunami maps on geomorphology, land use / land cover, tsunami inundation and finally the tsunami risk zonation map on 1:8,000 scale. In this process, water and soil samples of the affected regions were also collected and chemical analysis was carried out. The data on damages were collected using a specially formulated data sheet. High resolution IKONOS data of January 2000 and IRS data of January 2005 were used for the generation of respectively the pre and post-tsunami geomorphology and land use / land cover maps. Ground truth data were also incorporated in these thematic maps.

The study has revealed that the water in all the bore wells became brackish after tsunami. About 80 country boats, 40 fibre boats and nets were lost with an estimated loss of about one crore. The tidal waters damaged the bunds of the aqua-farms and deposited the transported sediments in them. The extensive casuarina plantations, were reduced by the onslaught of tsunami and the casuarina trees became brown due to their contact with tidal waters. From amongst the various geomorphic units mapped in the study area, the beach, spits, shoals, creeks, swales, palaeo creeks were inundated by tsunami where as beach ridges, dune complexes and stabilized sand dunes were not affected and hence inferred as barriers. Geomorphology, land use / land cover and the tsunami inundation maps, tsunami risk zonation map were prepared showing very high, high and moderate zones of tsunami risks. It is recommended to take up casuarina plantations extensively along the coast as a biowall, which will also provide employment to the local people. The shallow aquifers of the coastal environment were contaminated by the tidal waves. Hence, drinking water should be supplied through protected water supply schemes. As a part of the community development, proper awareness must be created through training and appraisal to the fisherman community to face such natural disasters. The coastal regulation zone act, which prohibits any construction activity within 500m of the high tide line, should be strictly enforced. Keeping in view of the experience, a holistic tsunami protection plan is needed to safe guard the Indian coasts.

Keywords: Geomorphology, Land use / land cover, Damage assessment, Nellore district, Tsunami Risk zonation.

INTRODUCTION

On 26th December 2004, an earthquake of 9.3 magnitude on Richter scale occurred off the west coast of Sumatra triggering a devastating tsunami in the Indian Ocean region. The Indian plate and Burmese plate beneath the Indian Ocean cracked past each other and the sea floor was forced upwards by some ten meters (Emma Marris 2005). These two plates move against one another at an average rate of about 6 centimeters per year, but this movement of 26, Dec 2004 did not happen smoothly (Michael Hopkin 2005). There has not been a very large quake along this fault since 1833 and this might have contributed to the huge force of this magnitude on that day. The India plate's jarring side released the tension on the Burma micro plate causing it to spring violently upwards. This displaced trillion tons of water, driving it towards Southeast Asia's coastline traveling at a speed of 900 kmph with larger wave length and less amplitude. When the waves reached shallower water near the coasts, it shortened, slowed and gathered into surges as tsunami that killed at least 1,50,000 people across a dozen South Asian countries. The earthquake, the world's biggest one during the last forty years and the fourth largest since 1900 has literally redrawn the map moving some islands by upto 20 meters. In the aftermath of tsunami, experts are welding together the details of the seismic slip that sparked the waves.

India had about 10,000 reported deaths, particularly in Andaman and Nicobar Islands, Tamil Nadu, Pondicherry, Andhra Pradesh and Kerala. In Andhra Pradesh about 630 km length of coastal region was affected covering six districts namely Nellore, Prakasam, Guntur, Krishna, West Godavari and East Godavari. Among them, the four southern coastal districts experienced maximum damage. The total length of the coast of these four districts is about 450 km with Nellore 165km, Prakasam 105km, Guntur 60km and Krishna 120km.

METHODOLOGY

The main objective of the study is to carry out tsunami disaster assessment by analyzing pre and post-tsunami geomorphology, land use / land cover, soil and groundwater quality using Remote Sensing, ground based data and GIS and to finally prepare a tsunami vulnerability map.

In the present paper, the work carried out in Tuppalapalem of Vagarru revenue village of Vakadu mandal in Nellore District (Fig.1) is discussed as a case study. The village, located in the southern part of the district, was damaged by the tsunami, due to its proximity to the shore. The village is also adjacent to the northern fringe of the Pulicat Lagoon. Tuppalapalem is included in the Survey of India toposheet No.66 B/4.

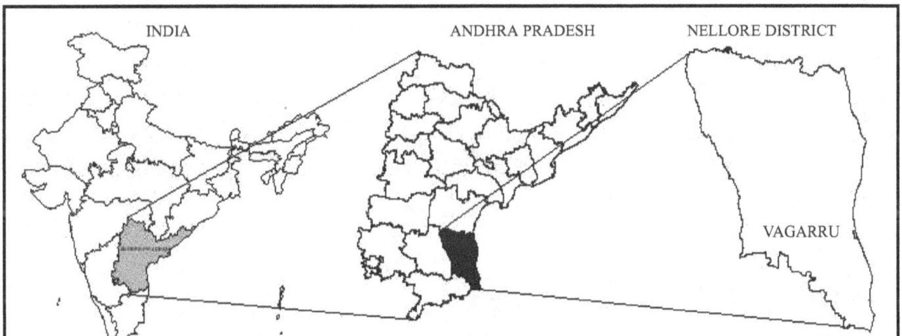

Figure 1: Location Map

The methodology involved the collection of information in four types of specially designed data formats on 1. Geology, water and soil 2. Structural and infrastructure damages 3. Changes in marine and coastal environment and 4. Land use / land cover changes. In order to understand the natural resource set up before and after tsunami, high resolution IKONOS data of January 2000 and IRS data of January 2005 were used respectively for the generation of geomorphology and land use / land cover maps on 1:8,000 scale. Tsunami inundation map was prepared using AWiFS satellite data of 27th December 2004 with duly supplemented by the preliminary field observations. Existing flood hazard map on 1:50,000 scale was also consulted. Chemical analyses were carried out for soils and waters collected from these affected areas. Further, corresponding GPS locations of water and soil samples and damaged sites were also collected. The available geological maps, published literature and other informations were used for enriching the thematic details. Ground truth verification was carried out in the form of well inventory, different types of damage to natural and physical resources. The same information was incorporated in the preparation of thematic maps. Keeping in view, of all these thematic information and chemical analyses data a tsunami risk zonation map was prepared for future planning strategies.

DAMAGE TO COASTAL ENVIRONMENT

The information collected in the above mentioned four data matrices for the villages were synthesized and presented as follows. There are about 500 houses in the Tuppalapalem village. No damage has been caused by the tsunami to these houses. There is no protected water supply scheme for this village and each house has an independent ground water source in the form of a filter point. It is reported that in a span of three hours i.e. from 8.30AM to 11.30AM, the tidal waves of tsunami struck this village 5-6 times. The water in all these filter points became brackish after tsunami. About 80 country boats, 40 fibre boats and nets were lost with an estimated loss of about one crore. The aquaculture is the major occupation for the people in this area. The tidal water damaged the bunds of these farms and deposited the transported marine sediments in them (Raghu et al 2005). The beach was having extensive coverage of casuarina plantation, which was substantially reduced by the onslaught of tsunami. The casuarina trees became brown after coming into contact with tidal water. A beach ridge was partially eroded uprooting the electric poles. The approach road for a length of about 300m to the beach was also damaged. There is no agricultural activity in this area as the livelihood of all the villagers is mostly fishing.

COASTAL GEOMORPHOLOGY

The synoptic view of satellite imagery facilitates better appreciation of geomorphology and helps in mapping the different landforms and their assemblage. The Survey of India toposheets on 1:25,000 scale provide topographic information required to interpret the geomorphology from the satellite imagery and thus different geomorphic units were mapped based on the landform characteristics, their aerial extent, thickness of deposition, etc. These geomorphic units / landforms were verified on the ground during the field visits and the ground truth information was incorporated and pre and post-tsunami geomorphological maps (Fig.2 & 3) of Tuppalapalem were prepared.

In Tuppalapalem, the coastal geomorphic units delineated are beach, spits, shoals, creeks, swales, beach ridges, palaeo creeks, dune complexes, stabilized sand dunes, etc, as detailed in table – 1

GEOMORPHIC SCENARIO OF VAGARRU VILLAGE, NELLORE DISTT. ANDHRA PRADESH

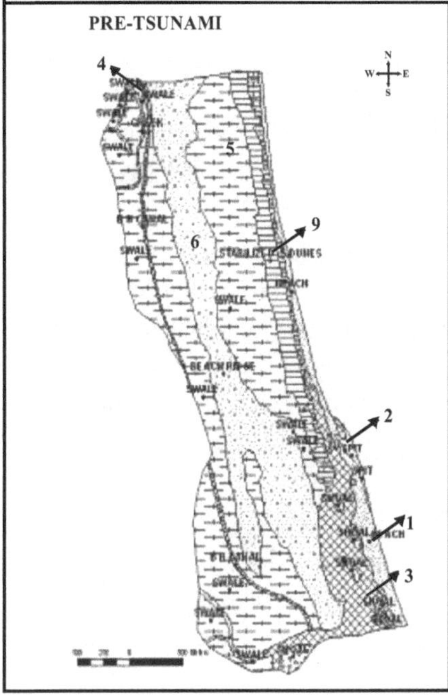

Figure 2 :

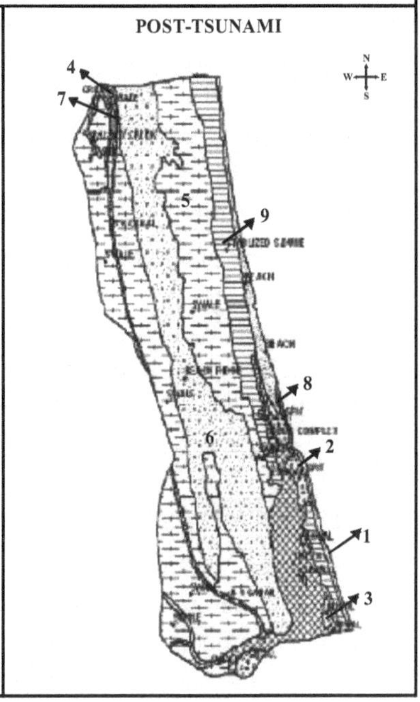

Figure 3 :

Table 1: Coastal geomorphology (ref Fig. 2 & 3)

S.No	Landform	Description	Remarks
1	Beach	Narrow strip of land along the seashore with unconsolidated sand / silt.	Persistent tidal action present
2	Spits	It is an elongated accumulation of sand as bar attached to the coast at one end and extending out to sea on the other end.	Persistent tidal action present
3	Shoal	It is a dynamic landform composed of non -clastic sediments emerges periodically	Persistent tidal action present
4	Creek	A linear water body connected to the sea	Persistent tidal action present
5	Swale	A linear depression occurring between two beach ridges parallel to the coast	Prone to inundation during flooding
6	Beach Ridge s	Sub parallel ridges of sand, silt with shell, normally parallel to the coastline withvarying amplitude	Acts as a barrier
7	Palaeo Creek	It is a linear depression with or without water, not connected to the sea	Prone for inundation during flooding.
8	Dune Complex	Colony of sand dunes generally of undulating surface near the shore	Acts as a barrier
9	Stabilized Sand Dune	It is a linear ridge formed by the deposition of wind blown sand near the shoreline covered with vegetation.	Acts as a barrier

On comparison of pre and post-tsunami geomorphology maps, significant changes were observed in the landforms. At the eastern side of the Tuppalapalem village, the Pulicat lagoon is joining the sea with a wide mouth. During pre-tsunami there was only one spit at the southern end of the said mouth. In a period of five years, a new spit has developed at the northern end of the Pulicat mouth, east of the village measuring to a length of about 500 m. Both spits are found converging towards the mouth of the lagoon.

But after the tsunami, a new shoal (400 m length, 100 m width) is observed on the southern side of the mouth constricting the width of the orifice of the lagoon. Few small shoals are also combined and formed a single shoal in the north of the orifice located to the west of the northern spit. The width of the beach on southern side of the mouth is reduced due to the development of casuarina plantations. While on the northern side, the width of the beach is increased to an extent of about 60 m.

LAND USE / LAND COVER

Remote sensing techniques offer benefits in the field of land use / land cover mapping and their change analysis. One of the major advantages of remote sensing is the repetitive coverage, which is necessary for change detection studies at global, regional, sub-regional and even cadastral scales.

The land use / land cover pattern of the district is varied as one travels from the mainland to the coast. The coastal villages are generally prevalent with shrimp farming running parallel to the coast and along the tidal creeks. Area from high tidal line (HTL) to the berm of the beach is covered with casuarina plantations, ranging from dense to sparse category. The tidal flats and swale regions are occupied with shrimp farms.

IKONOS MSS data of pre-tsunami was used for land use / land cover classification using ArcGIS 9 software, ARC/INFO and ERDAS IMAGINE. The pre-tsunami land use / land cover map was over laid on to the remote sensing data of post-tsunami period to demarcate the damages caused to various land use categories. The description of each unit is as follows.

Built-up Land

Vagarru village falls under rural category, wherein 80% of the population involved in agriculture based activities for their livelihood. All the agricultural villages covering 5 ha area and more are included under this category.

Plantations

It is described as an area under agricultural tree crops planted, adopting certain agricultural management techniques. It includes trees from mango and citrus family.

In Vagarru village *Acacia arabica* and casuarina are generally grown for fuel wood, in waste and sandy to marshy areas. Plantations are basically divided into forest and agricultural plantations, wherein forest plantations include Casuarina and those under agriculture base are casuarinas, which is mainly grown as commercial plants to supply wood to paper industry.

Aquaculture

Aquaculture is predominant along the areas surrounding Buckingham canal and Tidal Creeks. A few hatcheries are also found in this village.

Forest Plantations

Generally forest plantations are species of forestry importance and are raised on notified forest lands, whereas along the coastal tracts of this village casuarina plantations are grown so as to withstand high velocity winds, which accompany cyclones and other meteorological conditions.

Forest Blanks

These are the areas that are found as vacant gaps in the plantations which are generally caused either by cutting by humans or by natural degradation.

Land with Scrub

These areas generally fall into wasteland category, which are non productive and are generally used for grazing cattle or at times for fuel wood plantations.

Sandy Area

Coastal areas such as beaches, sand bars and shoals are included in this category.

Marshy Area

These are lands that are wet throughout the year or continuously flooded with surface water, not deeply submerged and are covered dominantly with sledges, bushes or other hydrophytes on the upland areas.

Canal / Tidal Creek /Water Logged Area

The lands, which are saturated with water either by surface congestion, flooding or high groundwater table are included in this category.

TSUNAMI DAMAGE ASSESSMENT

The changes in the land use / land cover were mapped and categorized by using pre and post-tsunami satellite data. It is inferred that the surge water did not cross the Buckingham canal which probably acted as a barrier for the in surging tsunami water. Much of the damage is seen along the coast and wherever creeks are found they acted as channels for the water to gush into the inland. Various land use / land cover units mapped in this village are given in the Table-2 and the pre and post-tsunami maps are shown in Figs.4 & 5. The total geographical area of the village is 794 hectares, but as the coastal tracts of the village has witnessed some erosional and depositional changes, an over all increase of 9 hectares was observed in the analysis.

Table 2: Tsunami induced damages in land use / land cover of Vagarru Village

Land use / Land cover units	Area in Hectares		Remarks
	Pre-Tsunami	Post-Tsunami	
Forest			
Forest Plantations	40	34	This unit was reduced by 6 hectares due to erosion near the creek.
Forest Blanks	22	55	This unit has increased as much of the area along the beach is devastated by the tsunami waves causing destruction of trees. The other reason is sand casting by the high tidal waves of the tsunami. An increase of 33 hectares is noticed.
Waste Lands			
Land with Scrub	123	165	This unit was increased by 42 hectares, as dense forest plantations, which were eroded by tsunami waves or by sand casting in between the trees.
Sandy Area	55	28	This unit was decreased as the creek connecting the sea was widened. Erosion is noti ced towards south of the creek and some of the stabilised sand bars also disappeared. Thus, there is a decrease of 27 hectares.
Marshy/ Swampy Land	156	168	This unit also has increased due to filling up of low lying areas such as swales, and other depressions where sea water pushed by the high rise tsunami waves. Much of the areas surrounding the creeks and canals were included into this category. 12 hectares increased in this unit.
Water Bodies			
Canals	49	49	There is no change in this land use category.
Tidal Creeks	73	92	The area under tidal creeks has increased which is attributed to the widening of the creek mouths by the intruding tsunami waters. An increase of 19 hectares was noticed.
Total Area	**518**	**591**	

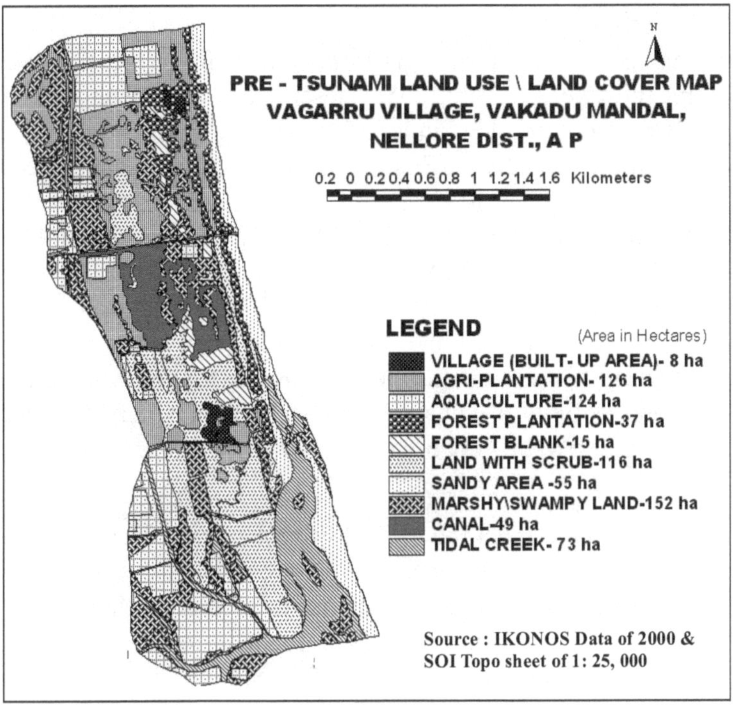

Figure 4:

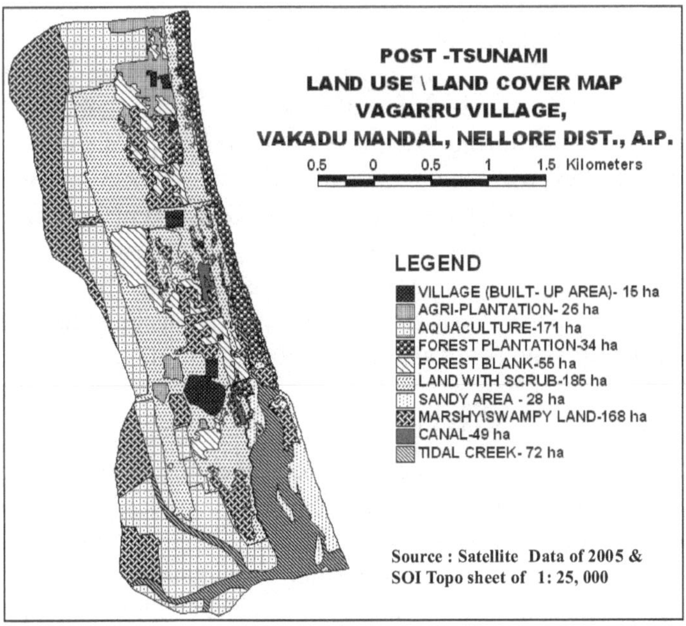

Figure 5:

CHEMICAL QUALITY

Water samples were collected in plastic containers from filter points in selected locations during third week of May 2005. Soil samples were collected from beaches. Chemical analysis was carried out in terms of pH, EC, TDS, Chlorides, Hardness, Alkalinity, Sulphates, Phosphates, Fluorides, Nitrates, Ca, Mg, Na & K for water and pH, EC, Alkalinity, Na, K & Ca for beach sand. The water analysis of Tuppalapalem revealed the pH of 8.5, EC 820, TDS 600, Chlorides 622, Hardness 450, Alkalinity 190, Fluorides 0.8, Nitrates <2 mg/l, Ca 320, Mg 130 and Na 1680. The pH of beach sand is 8.2, while the EC is 15420, Na is 44.6, K is 2, and Ca is 0.64. The EC is expressed in milli mohs/cm while the rest of the parameters are in mg/l. It is significant to note that the cations viz. Na, Ca and Mg are found to be very high in water sample of damaged wells. The anions namely chlorides and hardness are also found to be very high while the fluorides and nitrates are within the permissible limits.

The villagers reported that the ground water became brackish after tsunami, which is supported by very high concentration of cations (Na, Ca, Mg) and anions like chlorides and hardness. The sea water not only flooded the land surface but also intruded into ground water thus raising the ground water table and contaminating the water resources (Nageswara Rao et al 2005). Similar increase in salinity was observed in surface and ground waters and sediments in tsunami inundated areas in Thailand. The origin of these contaminants is attributed to marine as well as litho and anthropogenic nature (Witold Szczucinski et al 2005). The ground water quality can deteriorate in coastal areas due to direct inundation of land and slow seepage of salt laden seawater. There is an increase in EC in certain coastal tracts of Visakhapatnam, East Godavari and Nellore districts after tsunami (Prasad and Pradeep Raj 2005). They further concluded that the impact of tsunami on ground water regime is not severe. But, on the contrary, in many locations of the tsunami affected areas, the ground water became brackish after tsunami. However, the extent of contamination of ground water in this coastal region depends upon the nature of aquifer, type of well, nature of recharge and discharge rates. The observation wells of state groundwater department are located away from the study area and pre-tsunami water quality data is not available for comparison.

TSUNAMI RISK ZONATION

Tsunami risk map (Fig.6) was prepared based on the Flood / Cyclone Hazard Map (prepared earlier by APSRAC using SPOT, IRS and LANDSAT, on the basis of the drainage, slope, hydrology, geomorphology, history of past cyclones and floods) and coupled with recent tsunami inundation and field observations. The map was prepared to examine the occurrence and intensity of tsunami disaster and to assess the consequent risks in terms of hazards and vulnerability caused especially in the 1 km stretch along the coastal area of Nellore District of Andhra Pradesh and classified into Very high, High and Moderate zones of risks.

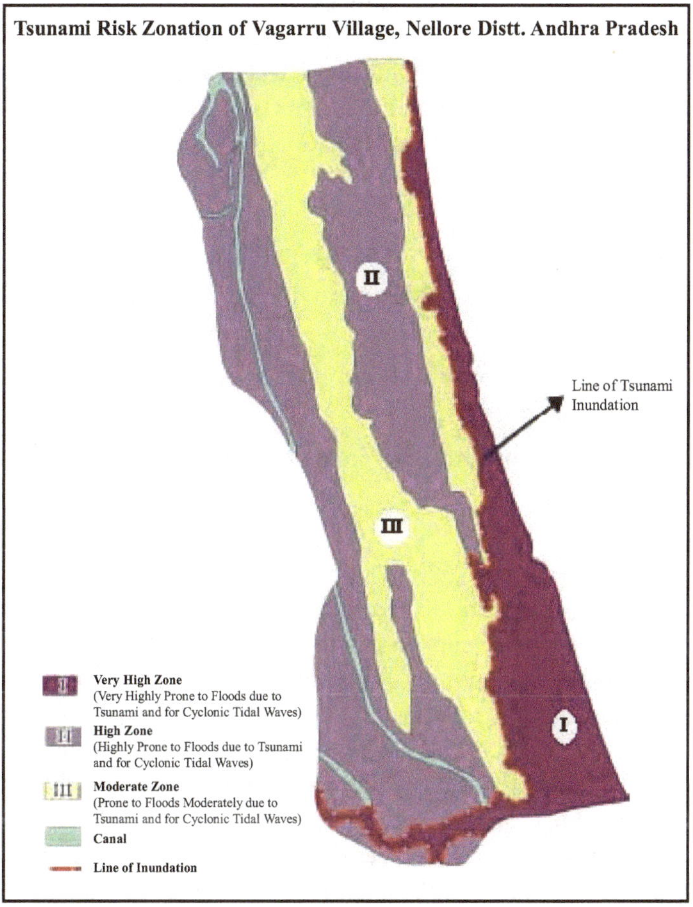

Figure 6:

The previous cyclone / flood hazard maps were having 13 classes based on the nature of submergence and inundation. The zonation of inundated areas with respect to the tsunami has lead to the modification of the above map. The zone – I boundary of the flood hazard map was altered and modified keeping in view of the inundated line prepared from the recent tsunami. Much of the area in zone – II of flood hazard map has now fallen in the zone – I of risk map of the tsunami, because the killer tidal waves have encroached much towards inland side relatively. The surge water due to extraordinary height of tsunami waves have reached further inland where creeks, out falls and river mouths are present. In some of the areas under study, it was observed that severe beach erosion has occurred due to high velocity waves through which tsunami tidal waves have reached much inland side there by causing severe damage. Based on this information on present tsunami inundation, a tsunami risk map was prepared. The area was classified into very high risk zone – the area very highly prone to flood due to tsunami and /or cyclonic tidal water / storm surge and heavy rainfall (in situ / catchments), high risk zone – the area highly prone to flood due to tsunami and /or cyclonic tidal water and heavy rainfall (in situ / catchments) for longer duration and moderate risk zone - which is prone to flood moderately due to tsunami and /or cyclonic tidal water and heavy rainfall (in situ / catchment) for shorter duration.

Tsunami Risk Zonation of Vagarru Village

i) ***Very High Zone:*** Very highly prone to flood due to tsunami and / or cyclonic tidal water

ii) ***High Zone:*** Highly prone to flood due to tsunami and / or cyclonic tidal water

iii) ***Moderate Zone:*** Prone to flood moderately due to tsunami and / or cyclonic tidal water

CONCLUSIONS

In the habitations close to the coast, fishing boats and nets were extensively damaged. No protected water supply scheme is present in majority of the villages for drinking. The tidal water has deposited sediments in aqua farms damaging the bunds. Casuarina plantation was destroyed and became brown after coming into contact with saline water. Limited available drinking water sources were contaminated with saline water. Low lying areas were inundated with surge water and stagnated with huge amount of saline water for a couple of days creating unhygienic conditions associated with epidemic diseases. Shalabh Bharadwaj et al (2005) using low resolution MODIS satellite images of pre and post-tsunami event observed that the Duggarajupatnam area of Nellore District falls in very high risk zone as the chances of epidemic break is very high due to inundation and stagnation of tidal water. This village is adjoining the southern boundary of present study area.

Relative disasters depend on the coastal zone geology comprising lithology, active tectonics and micro level geomorphological conditions, geometry of the coasts etc. The geometry is controlled by tectonics and geomorphology. The convex coasts are the emerging coasts due to cymatogenic arching and block faulting and are comparatively less prone while the concave coasts are the subsiding coasts due to cymatogenic deeps and grabens and hence are more prone to tsunamis and storm surges (Ramasamy et al 2005). The study area is located on the concave part of the Nellore coastline. Efforts are to be made to protect, rehabilitate and enhance the livelihoods of the coastal communities in a suitable manner with a better preparedness plan. It is reported that beach ridges, and casuarina plantations have acted as barriers and reduced the onslaught of tsunami. The bunds and embankments of the creeks should be strengthened. It is recommended to take up casuarina plantations extensively along the coast as a biowall, which can also provide employment to the local people. The shallow aquifers of the coastal environment were contaminated by the tidal waves. Hence, drinking water should be supplied through protected water supply scheme. As a part of the community development, proper awareness must be created through trainings and appraisal to the fishermen community to face such natural disasters. Tsunami hazard perception by the people of the coastal area is necessary at least at basic level in mitigating the loss of life and property (Mruthyunjaya Reddy et. al 2005). The coastal regulation zone act, which prohibits any construction activity with in 500 m. of the high tide line, should be strictly enforced. Keeping in view of the coastal geologic, morphotectonic setting and land use / land cover a holistic tsunami protection plan is needed to safe guard the Indian coast. These plans shall try to locate tsunamigenic earthquakes in near real time, deploy ocean bottom sensors to detect generation of tsunamis, calculate the travel time from possible sources to Indian coasts based on bathymetry and other relevant information, prepare inundation maps and the effect on coastal areas and develop a proper mechanism of communication as early as possible to the stakeholders.

ACKNOWLEDGEMENT

Grateful thanks are due to the Department of Science and Technology, Govt. of India for sponsoring the project on tsunami damage assessment and mapping of tsunami affected areas in four districts of Andhra Pradesh. The authors are greatly indebted to Brig. Dr. R.Sivakumar, Head, NRDMS and Dr.Bhoop Singh, Director, Department of Science and Technology, Govt. of India for their constant encouragement and support.

REFERENCES

Emma Marris (2005) Inadequate warning system left Asia at the mercy of tsunami. Nature, v. 433(6), 3p.

Michael Hopkin (2005) Triple slip of tectonic plates caused seafloor surge. Nature, v. 433(6), 3p.

Mruthyunjaya Reddy, K., Nageswara Rao, A and Subba Rao, A.V (2005) Recent tsunami and its impact on coastal areas of Andhra Pradesh. In: Tsunami: The Indian Context, (eds: SM.Ramasamy and C.J. Kumanan), Allied Publishers Pvt. Ltd., New Delhi, pp. 129-138.

Nageswara Rao, A., Raghu, V., Subba Rao, A.V. and Mruthyunjaya Reddy, K (2005) Impact of tsunami on waters and soils along the coast of Prakasam District, Andhra Pradesh (abs.). In: National Seminar on GIS Application in Rural Development, (eds.

V.Madhava Rao, R.R. Hermon, P. Keshava Rao), National Institute of Rural Development, Hyderabad, pp. 326-330.

Prasad, B.N. and Pradeep Raj (2005) A note on effect of tsunami on ground water regime in parts of Andhra Pradesh. Journal of the Geological Society of India, v. 66, pp. 277-280.

Raghu, V., Nageswara Rao, A., Subba Rao, A.V. and Mruthyunjaya Reddy, K. (2005) Damage to coastal environment due to 26th December 2004 tsunami in Nellore District, Andhra Pradesh (abs.). In: National Seminar on GIS Application in Rural Development, (eds. V.Madhava Rao, R.R. Hermon, P. Keshava Rao), National Institute of Rural Development, Hyderabad, pp.341-344.

Ramasamy, SM, Kumanan,C.J., Selvakumar,R. and Saravanavel, J. (2005) Seismo-tsunamis scenario in India, GIS@development, v. 9, pp. 18-20.

Shalabh Bharadwaj, Siva Subramanian, K.S., Taniya Ray, and Sandeep Gupta (2005) Mapping epidemic prone areas. Geospatial Today, v. 4(3), pp. 20-25.

Witold Szczucinski, Przemyslaw Niedzielski, Grzegorz Rachiewicz,Tadeusz Sobczynski, Anetta Ziola, Artur Kowalski, Stanislaw Lorenc, and Jerzy Siepak (2005) Contamination of tsunami sediments in a coastal zone inundated by the 26 December 2004 tsunami in Thailand. Environmental Geology, v. 49(2), pp. 321-331.

GEOMATICS IN TSUNAMI
S.M. Ramasamy, C.J. Kumanan, Brig. R. Sivakumar and Bhoop Singh (eds.), pp. 77-85
New India Publishing Agency, New Delhi (India)
E-mail : spjain_niph@rediffmail.com Web: www.bookfactoryindia.com

7

TSUNAMI – 2004 DAMAGE ASSESSMENT MAPPING, CHENNAI REGION

M.Ramalingam, M.Kamaraj, K.Vaithiyanathan and RM.Narayanan

Abstract

The recent tsunami occurred on 26th December 2004, caused great devastation along the east coast of Tamil Nadu, India. As the said tsunami has caused colossal damages, the impact assessment has become essential for tsunami mitigation measures. In this context, the extent of tsunami inundation, loss of lives, property loss and damages to physical and natural resources have been assessed in the tsunami affected area between Ennore and Mahabalipuram along the east coast of Tamil Nadu.

IRS LISS III PAN merged satellite data and IKONOS data were used for mapping the pre-tsunami geomorphology and land cover conditions and IRS-P6 PAN merged satellite data were used to map the post – tsunami terrain conditions including the inundations. From the study it is inferred that wherever buildings and population are denser the loss is more. Hence, vertical expansion may be needed as an alternative or to supplement horizontal expansion. Bio-shield along coastal areas, involving raising of mangrove forests, planting of casurina, salicornia, palms, bamboo and other tree species and halophytes which can grow near the sea will serve as speed-breakers for the up rising waves.

Keywords: Tsunami, Damage Assessment, Mitigation Measures, Chennai region.

INTRODUCTION

Mapping of damage assessment is one of the most significant applications of satellite remote sensing. Near real time monitoring and damage assessment has been successfully operationalised now a days in India.

On 26th December 2004, tsunami waves struck the east coast of Tamil Nadu, Andhra Pradesh and Andaman - Nicobar Islands of India, Sri Lanka, Indonesia, Malaysia, Burma, Thailand, Maldives, East African coast, etc. This tsunami was triggered by the subduction of Indian plate under Burma micro - plate along a N–S fracture zone extending along Andaman – Nicobar Islands and the resultant massive subterranean earthquake with magnitude of 9 M Richter scale at northern Sumatra.

The killer tsunami waves that struck with 10 -12 m height at 9.00 am in the morning of December 26, 2004 played havoc on the Ennore beach, Kasimedu fishing harbour, Marina beach, Srinivasapuram - Foreshore estate, Besant Nagar beach and several fishermen villages along the east coast, engulfing human lives, fisher man huts and houses, fishing nets, boats, vehicles, roads, trees and plantations. The devastating effect was felt more on the marina and Foreshore estate with heavy damages to building and human lives.

Department of Science and Technology, Government of India has entrusted the work of assessing the extent of tsunami inundation, loss of lives, property loss and damage assessment for Physical and Natural Resources in the tsunami affected area between Ennore and Mahabaliburam to Institute of Remote Sensing, Anna University.

Study Area

The study area is the coastal stretch extending from Ennore to Mahabaliburam along the east coast of Tamil Nadu. The total length of the study area is about 80 km and the width is about 1 km and is covered in the SOI toposheets of 66C/8, 66D/1 and 66D/2. The Ennore Port, Chennai fishing Harbour and Chennai Port are located along this coastal stretch. The beautiful and world's second largest 'Marina' beach is located in this zone. Several numbers of fishing villages, parks, restaurants and beach resorts are also located in this stretch (Fig.1).

Objectives

The main objectives of the study are to understand the impact of tsunami over natural resources, physical resources and human resources:

➢ Natural Resources

- Determination of Soil salinity and land use / land cover changes, assessment of contamination of potable ground water resources along the study area and Preparation of suitable plan

➢ Physical Resources

- Preparation of risk maps for future planning strategies

➢ Human Resources

- Loss of life

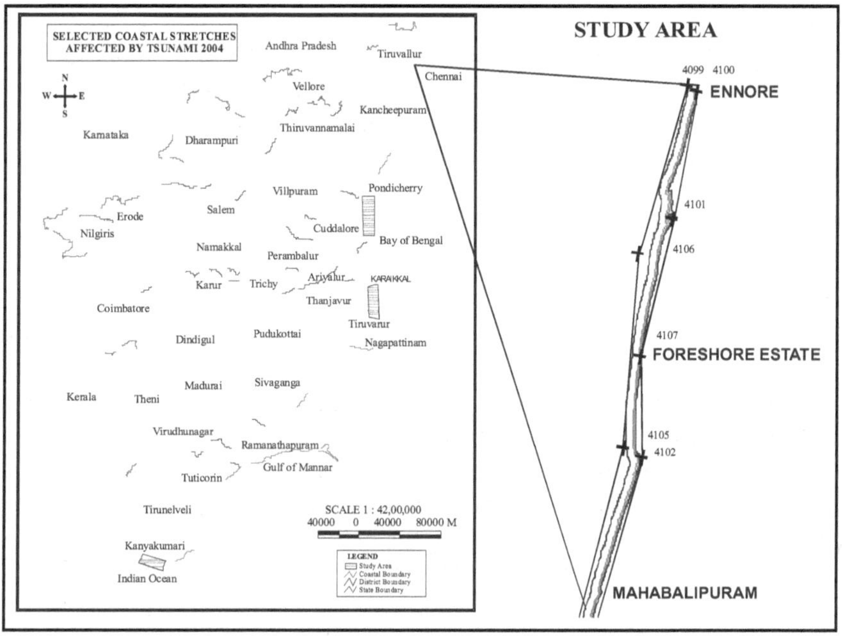

Figure 1: Study area

METHODOLOGY

Using Remote Sensing and GIS technology, the pre-tsunami natural resources maps have been prepared with the help of SOI Toposheets (1: 25, 000 Scale) and Cadastral maps (1: 8,000 Scale). LISS III PAN merged satellite data, IRS-P6 PAN merged satellite data and IKONOS satellite data were used for the preparation of Geomorphology, land use / land cover and tsunami inundation maps.

Extensive field investigation was undertaken after the tsunami in the study area. Which includes soil and water sampling for quality assessment in the tsunami inundated areas and evaluation of damages in affected fishermen villages along the coast and the results obtained are presented in the article.

REGIONAL GEOLOGY

The coastal zone between Ennore and Mahabalipuram comprises recent coastal Alluvium, with an admixture of fine-grained light white sands with broken shell pieces. In the creeks and low lying areas away from the sand ridges to the west, dark clays enriched with salts are extensively found along with dead shells. The coastal sands lie over the Charnockite basement with an average thickness of 12-20 m and intercalated with occasional reddish brown clays.

Ground water occurs in the beach environment under perched conditions at shallower depth with an average depth of 3 to 4 m and the fresh water is collected from monsoonal rains. The 'Tiuvanmiyur aquifer', a fresh water aquifer, is found between Thiruvanmiyur and Muttukadu with a vast reserve and supplying water to south Chennai areas. The aquifer thickness is varying between 12 and 20 m. The greater exploitation of this aquifer for water supply by various agencies often results in the saline water intrusion in certain areas like Besant Nagar. In creeks, salt pans and along Buckingham canal, water is always saline due to the sea water coming into them during high tides.

GEOMORPHOLOGY

Geomorphologically, the area comprises younger and older coastal alluvial plains. The younger coastal plain is characterised by narrow to wider beaches followed by beach and sand ridges which are gently sloping towards sea side and swale complexes of swamps / marshes, mud flats and salt flats. The salt flats are utilized as saltpans around Muttukadu. The creeks receive seawater during high tides. The East Coast Road (ECR) is running along the beach ridges and the settlements and resorts are located along the high way (Fig.2).

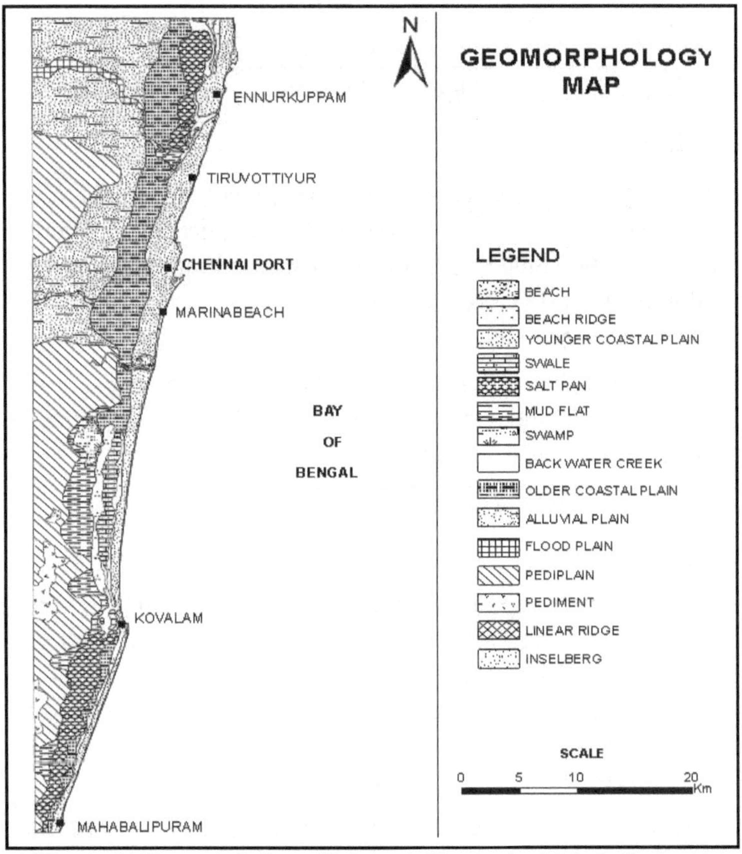

Figure 2: Geomorphology (Ennore – Chennai – Mahabalipuram)

LAND USE /LAND COVER

The land use classes present in the study area is shown in Fig.3

Agricultural lands	:	Crops Lands, Plantations, orchards, garden plants and Fallow Lands
Forestlands	:	Land with Scrub, Land without scrub, Forest Plantations and Mangroves
Water bodies	:	Water logged areas, Creeks and ocean
Wastelands	:	Salt affected lands, Marshy/swampy lands, Saltpans, Mud Flats, Beaches and Sandy Area,
Settlements	:	Towns/Cities, Villages, Industries and road net works

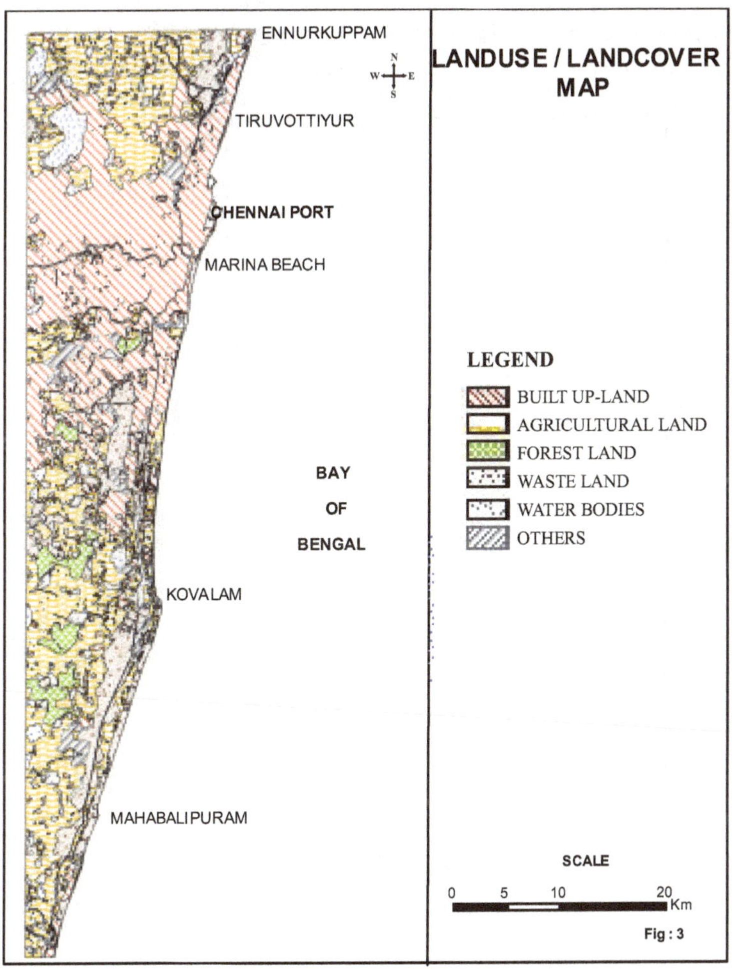

Figure 3: Land use /Land cover (Ennore – Chennai – Mahabalipuram)

DISASTER ASSESSMENT

Natural Resources

For the natural resources disaster assessment, surface soil sampling and ground water sampling were done to assess the extent of damage to the soil and sub soil stratum and the contamination level of the coastal aquifer due to the invasion of the seawater during the tsunami episode.

Soil Salinity

The soil samples have been collected from 7 selected locations and analysed. The results indicate that the soils in the study area are neutral in reaction except sample no.S3, which is moderately alkaline with pH of 8.4. The samples S3 and S6 have medium Salinity when compared to other locations. This can be reclaimed through the monsoon rainfall and frequent flushing. The TDS map of soil is shown in Fig.4 for example.

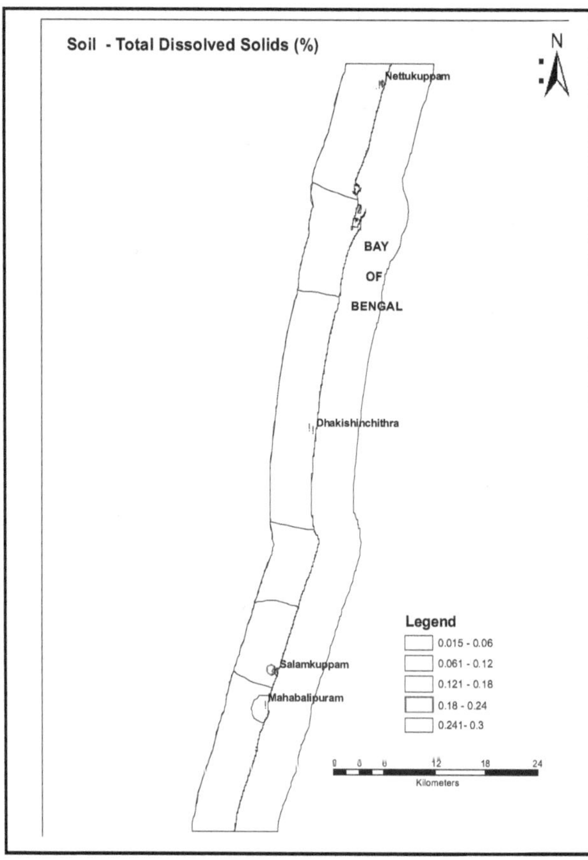

Figure 4: TDS (%) of soil

Groundwater

The Groundwater samples were collected from 40 locations all along the coast of the study area and analysed for possible contamination, if any. The water quality results show that the following villages have been affected by the tsunami such as Nochikuppam, Panaiyur, Tiruvottiyur, Salankuppam and Periyakuppam. The pH ranges from 6.59 to 8. The Total Dissolved Solids ranges from 86 m/l to 25,000 m/ l and the maximum values are found to occur in Devaneri, Vadanemmeli, Idayar, Nochikuppam, Nemmelikuppam and Semmancherikuppam villages. In these areas, the perched aquifers have been contaminated with seawater during the tsunami invasion and became non-potable and these aquifers can be improved by the frequent flushing of rainwater. Fig.5 shows the TDS variations in ground water in the study area, which may be due to tsunami. Due to lack of historic data these results could not be compared with the pre-tsunami data.

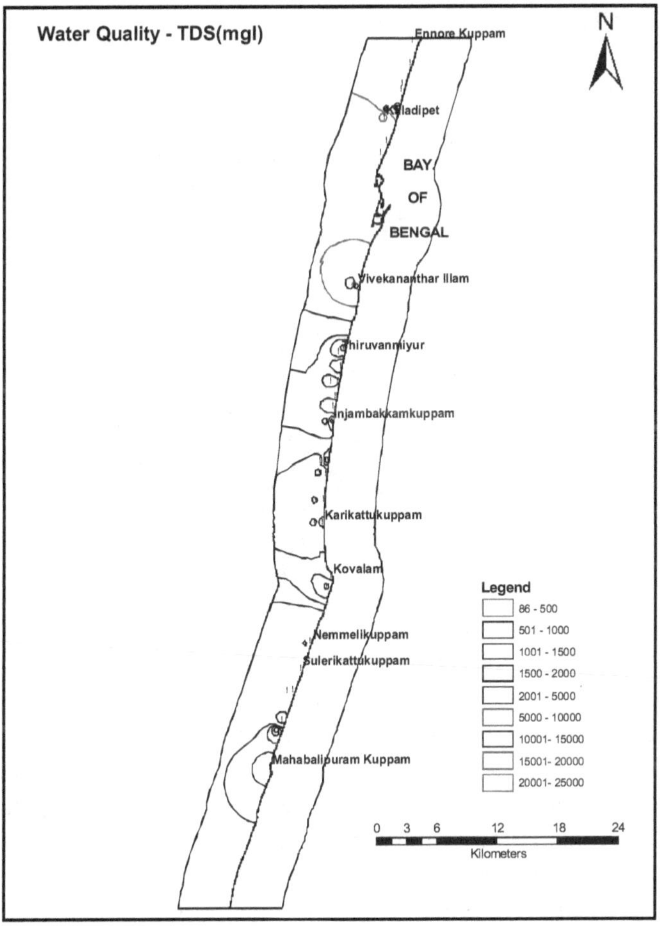

Figure 5: TDS of groundwater

Physical Resources

Physical Resources assessment have been made from Ennore to Mahabalipuram, to understand the coastal erosion, deposition and damages to physical structures – Buildings, Houses, Huts, Roads, Fishing Boats etc. The field observations indicate that tsunami waves have forcefully entered 100 – 250m in the coastal areas with more inundation of 1 – 2 km in the creeks and rivers of Adyar and Coovam and breached the dune ridges of Adyar Mouth Sand Bar, Coovam Sand Bar and Muttukad creek. The unconsolidated sandy materials along the coast have been eroded and redeposited during the backwash. The tsunami has made the beach undulating with the formation of trenches and pits in several places and filling the deposits in low-lying areas.

Figure 6: IKONOS Satellite Data

The study revealed that the structural damages are in the form of 1163 numbers of RCC buildings, 142 numbers of madras roofed buildings, 70 numbers of mangalore tiled houses, 169 numbers of Asbestos ones and 2916 numbers of huts, 4097 numbers of boats and 5543 numbers of fishing nets. Almost all of the fisherman huts located in the study area were washed away along with the materials, utensils etc. Boats in the fishing harbor were shattered and thrownoff far away. Several cars parked in the Marina beach were thrown out and severely damaged. The physical damages observed in the Kasimedu has been mapped using IRS P6, LISS- IV satellite data and IKONOS images for pre and post-tsunami, respectively and the inundation line delineated from the IKONOS images is shown in Fig.6.

Human Resources

This tsunami with catastrophic force caused disaster and loss of lives in some of the coastal villages. Over 70,000 people including fisherman community and the people taking morning walk on that fatal Sunday (26, December 2004) along the beach were watching the tsunami without knowing what was happening. Nearly 200 people have lost their lives. Apart from that, animals have also lost their lives.

DISCUSSIONS AND CONCLUSIONS

The coastal stretch extending between Ennore and Mahabalipuram was studied for assessing the damages occurred due to tsunami, which struck the Indian coast on 26th December 2004. LISS III PAN merged satellite data and IKONOS data (pre-tsunami), IRS P6, LISS- IV satellite data (post-tsunami) were used for assessing the natural and physical resources damages due to tsunami. The detailed field investigations included, water sampling, soil sampling, assessment of damage to the property etc.

In the coastal areas, where building and population densities were high, with roads, bridges and other buildings vastly spread over horizontally suffered to the maximum. Such areas may require vertical developments with identified horizontal pathways for speedy evacuation of people during the crises.

Creation of bio-shield along coastal areas, involving the raising of mangrove forests, planting of casurina, salicornia, palms, bamboo and other tree species and halophytes, all of which can grow near the sea will serve as speed-breakers under conditions of coastal storms, cyclones and tsunami.

The construction of permanent sea walls can be taken up in places wherever there is sea erosion due to heavy anthropogenic pressures. The locations for such non-living barriers should be determined on the basis of a carefully conducted erosion vulnerability analysis. This can be applied to the Tiruvotriyur coastal area.

Vulnerability mapping can be made based on the analysis of 100-years data and the areas prone to cyclones and other natural disasters can be mapped. Priority may be given to such areas in erecting bio-shields and in undertaking eco-restoration and erosion prevention measures.

Scientific land and water use planning is warranted in this area due alarming salt water intrusion problems.

GEOMATICS IN TSUNAMI
S.M. Ramasamy, C.J. Kumanan, Brig. R. Sivakumar and Bhoop Singh (eds.), pp. 87-106
New India Publishing Agency, New Delhi (India)
E-mail : spjain_niph@rediffmail.com Web: www.bookfactoryindia.com

8

DAMAGE ASSESSMENT IN THE TSUNAMI AFFECTED PONDICHERRY – CUDDALORE COAST, INDIA

G.V. Rajamanickam, E. Saranathan, V. Rajeshkumar, Sujatha E Ramani, D. Soosai Manickaraj, R. Chandrasekaran, K. Muthumeena, R. Mahesh, C. Sriramkumar, A. Rajarajeswari and V. Sathiskumar

Abstract

The tsunami has struck many coastal countries of South Asia on 26 December 2004 and caused greater devastation along Andaman - Nicobar and Indian coasts. In India, Tamil Nadu coast is worst affected in Kanyakumari, Nagapattinam and Cuddalore districts. Subsequently, damage assessment, in respect of natural and physical resources, was done using Remote Sensing and GIS in parts of Cuddalore and Pondicherry coasts. The structural damages were studied village wise and categorized as maximum, moderate and least affected areas. As far as the natural resources are concerned, the changes in geomorphology, land use/ land cover, soil and water have been studied. On the basis of the same, the risk maps were also generated in GIS environment.

Keywords: Tsunami, Damage Assessment, Remote Sensing, GIS, Cuddalore, Pondicherry

INTRODUCTION

The wonder struck surprising 26th December 2004 tsunami warranted detailed assessment of different damages in the area proposed for the study, i.e., Pondicherry to Coleroon coastal stretch on the shores of the Bay of Bengal between North latitudes 11°27' and 12°00' & East longitudes 79°50' and 79°58'. It is interesting to look into this area being the typical zone manifested by three adjacent canyons in the vicinity of the innershelf of this region. Further, the region under study is fairly populated having minor harbours like Pondicherry, Cuddalore and a fishery harbour in Coleroon mouth. Above all, the distinct feature of this region is the presence of the second largest mangroves forest in this country at Pichavaram. The study area covers mainly the delta portion of Cauvery including the Veeranam irrigated area. Major rivers of Tamil Nadu like Coleroon and Vellar are draining this region. In addition, particularly the Cuddalore region is found to be carpeted by dunes of both transverse and parabolic morphology. In view of many distinct physiographical coastal features, the understanding of tsunami impact in these varied landforms is expected to throw much deeper picture on the interaction of the landforms with the onward rushing tsunami waves.

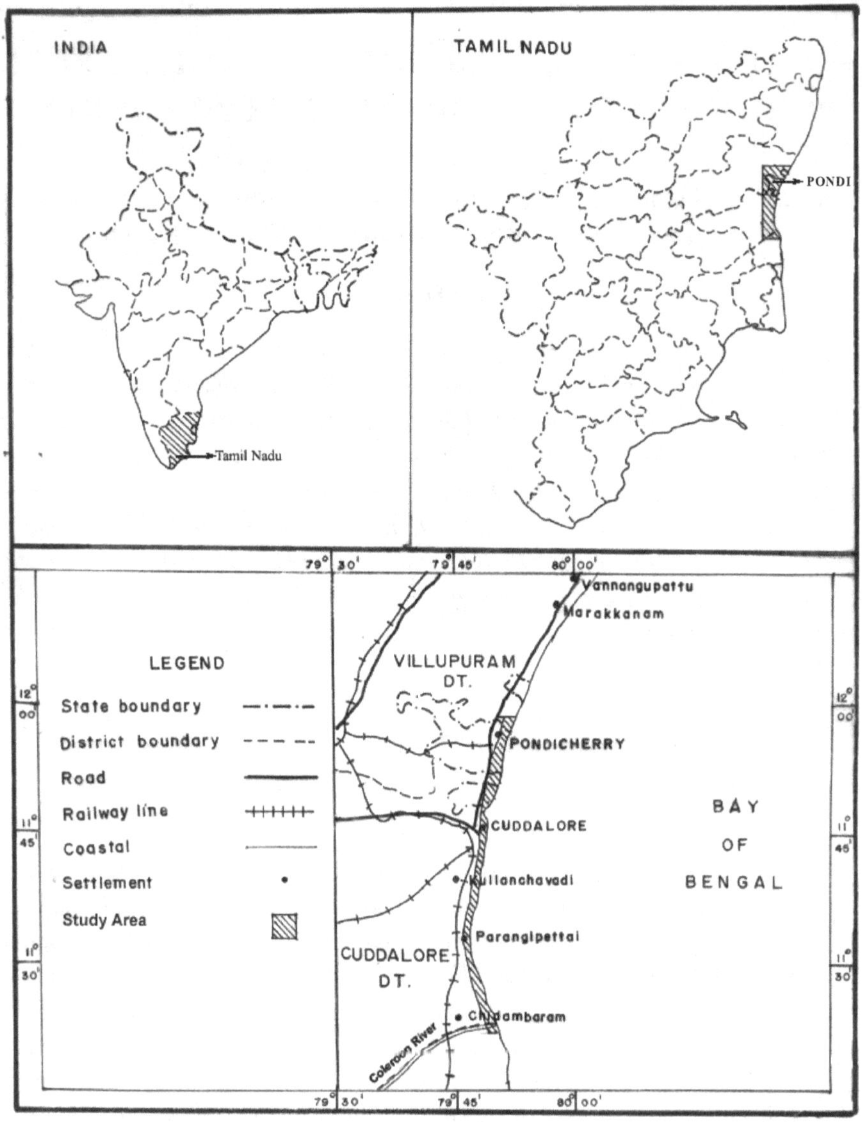

Figure 1: Study Area

To achieve this aim, the following objectives were kept in the cards:

- Preparation of damage assessment maps on Natural, Human and Physical Resources in GIS environment on 1:8,000 scale.
- Determination of soil salinity and land cover changes including erosion and deposition profile of the affected area.
- Assessment of contamination in drinking water.
- Preparation of risk map for future planning strategies and sustainable plans.

METHODOLOGY

The primary data such as toposheets, cadastral maps, satellite images and groundwater details were collected from various Government and Non-Government sources. The socio-economic aspects (loss of life, houses, boats, paddy fields, groundwater wells, potable water wells, road networks etc.) and details on civil constructions were collected by census method of door-to-door and person-to-person contact and direct interviews.

The land use changes, geomorphological changes and damages in the structures due to tsunami were noted down during the field visits. Field visits have been arranged for 140 days in such a way to cover the length and breadth of the entire area.

The thematic maps such as Base map, Urban/Village location, Land Use (pre and post-tsunami), Geomorphology, Drainage and Structural Damages maps were prepared in the scale of 1:8,000. The thematic maps prepared were digitized, edited for errors and linked with attributes with the aid of GIS software.

The multi-criteria analysis of the database was created adopting the following steps: (i) Preparation of Structural damage assessment map (Fig.2). (ii) Preparation of Erosion and Deposition map along the coast line (Fig.3). (iii) Mapping of land use changes (Fig.8a & 8b). (iv) Preparation of soil salinity and ground water changes map, preparation of sustainable plan for the affected region and (v) Preparation of Risk Map by the evaluation of Damage Assessment using the results of the above said analysis (Fig.10).

The results obtained are analyzed with reference to the Structural damages, Geomorphological changes, Land use changes and Hydrogeological deterioration. Finally, using the comprehensive data, a risk map was attempted.

STRUCTURAL DAMAGES

The Structural damages in Cuddalore, Chidambaram and Pondicherry regions due to tsunami are assessed and classified in 3 groups according to the percentage of destruction, i.e., 25 to 50 %, 50 to 75 % and >75 %.

Cuddalore Taluk

Out of 10 revenue villages in this taluk, the maximum damage is noticed in the Cuddalore municipality region (535 houses in the level of>75 %, 230 in 50 to 75 % and 153 in 25 to 50 %) and Gundu Uppalavadi (170 houses in the level of>75%, 50 in 50 to 75 % and 50 in 25 to 50 %). The reason for this extent of damage is presumed to be that the areas are densely populated, low lying (about 1m above MSL) and very near to the Coast (about 100 m). Even though the villages such as Kudikadu (23 houses in the level of>75 %, 14 in 50 to 75 % and 34 in 25 to 50 %), Pachaiyankuppam (29 houses in the level of>75 %, 8 in 50 to 75 % and 35 in 25 to 50 %), Kayalpattu (30 houses in the level of>75 %, 7 in 50 to 75 % and 06 in 25 to 50 %), Thiruchopuram (15 houses in the level of>75%, 11 in 50 to 75 % and 50 in 25 to 50 %), Andarmullipallam (25 houses in the level of>75 %, 17 in 50 to 75 % and 75 in 25 to 50 %) and Thiyagavalli (44 houses in the level of>75 %, 6 in 50 to 75 % and 15 in 25 to 50 %) are found very close to the Coast (about 100 m).In spite of that, the structural damages are observed to be less as they are at a higher elevation (about 3.0 m above MSL) compared to the earlier referred villages in this taluk. The damage in the village Suba Uppalavadi (2 houses in the level of 25 to 50 %) is very minimum, though it is found adjacent to the Gundu Uppalavadi, much less damage is attributed to the presence of Sand dune of about 3 m parallel to the Coast and the location of the settlement being comparatively away (about 250 m) from the Coast. The damage found in Madalapattu is nil since the settlement located is far away (about 1.5 km) from the Coast.

Photo 1: Damage of 25 – 50 % - Ariyagoshti

Chidambaram Taluk

Out of 10 revenue villages in this taluk, Killai village is found to be worst affected in such a way that the hamlets namely Pillumedu (120 houses), Kannagi nagar (12 houses) and Chinnavaikkal (58 houses) are noted to have been washed off by tsunami waves. All the structures seen in Hamlet M.G.R.Thittu (128 houses in the level of >75 %, 39 in 50 to 75 % and 10 in 25 to 50 %) are damaged to the level of that no body can use the structures further. Hamlets Mulukku thurai (50 houses in the level of >75 %, 37 in 50 to 75 % and 63 in 25 to 50 %) and Vadakku Murasalodai (95 houses in the level of >75 % and 20 in 50 to 75 %) belonging to the same revenue village are also affected severely. Even though many sand dunes are (at about 3 m from MSL) found along the Coast of these villages, the reasons for this worst damage are assigned to the location of the village having been bounded by Picharavaram backwaters to the South and west, Bay of Bengal to the East and Vellar River mouth to the North. On rushing tsunami waves have entered these villages situated behind the dune, through the backwaters. They are being in lower elevation than dunes keeping the dunes as bank, the marine water surrounded these villages. The revenue village Ariyagoshti (265 houses in the level of >75 %, 371 in 50 to 75 % and 278 in 25 to 50 %) is mostly affected by tsunami waves. This village is located at about 300 m from the Coast and at an elevation of about 1 m from MSL.

Photo 2: Damage of 50-75 % Villiyanallur

Photo 3: Damage of >75 % M.G.R. Thittu

The reason behind this, is that the tsunami waves are conduited by the sand dunes present around the Killai village, flooded this whole village (to about 2 m height). The damages in the other villages such as Parangipettai (71 houses in the level of >75 %, 40 in 50 to 75 % and 359 in 25 to 50 %), Kottattai (75 houses in the level of >75 %, 250 in 50 to 75 % and 28 in 25 to 50 %), Silambimangalam (56 houses in the level of >75 %, 76 in 50 to 75 % and 210 in 25 to 50 %), Villiyanallur (60 houses in the level of >75 %, 85 in 50 to 75 % and 48 in 25 to 50 %) and Periyapattu (46 houses in the level of >75 %, 29 in 50 to 75 % and 127 in 25 to 50 %) are noticed to be moderate. The damage in the village Kil thirukalipalai (11 houses to the level of 25 to 50 %) is seen minimum since it is located at a distance of 1.5 km from the Coast. The flooding in this village has occurred from the Kollidam River, which is located to the south of this village. The damage assessed is nil in the villages Pichavaram (as located 4 km away from the Coast) and Thandavaraya Cholagan pettai (as located 2.5 km from Coast). The main reason for the lack of damage is the presence of Mangrove shield in the Pichavaram backwater between the settlements and the Coast.

Pondicherry Region

This region consists of 10 revenue villages. The settlement in the region is seen denser along the Coast. The assessment in this region shows maximum damage in the villages namely Pillaichavadi (232 houses in the level of>75 %, 35 in 50 to 75 % and 7 in 25 to 50 %), Kalapet (246 houses in the level of>75 %, 57 in 50 to 75 % and 12 in 25 to 50 %), Ariyankuppam (110 houses in the level of>75 %, 73 in 50 to 75 % and 8 in 25 to 50 %) and Poornankuppam (129 houses in the level of >75 %, 76 in 50 to 75 % and 22 in 25 to 50 %). All the above villages are located at about 100 m from the Coast and at an elevation of about 2 m from MSL.The villages Pillayarkuppam (51 houses in the level of>75 %,35 in 50 to 75 % and 13 in 25 to 50 %), Kirumampakkam (48 houses in the level of>75%, 60 in 50 to 75 % and 210 in 25 to 50 %) and Manapattu (45 houses in the level of>75%, 23 in 50 to 75 % and 1 in 25 to 50%) are moderately affected by the tsunami waves. These villages are located at a distance of about 100 m from the Coast and at an elevation of about 3 m. The damage in the Pondicherry Municipality was assessed to be minimum (27 houses in the level of>75%, 2 in 50 to 75 % and 7 in 25 to 50%). In the main part of the Municipality, no damage was noticed where the Coastal stretch is well elevated at about 4 m from MSL and is protected with the help of boulders laid along the Coast. The damage is seen only in the outskirts of the municipality where, it is not protected by boulders. The damages assessed is nil in the villages Thengatittu and Manavely. Backwaters bound the village Thengatittu on three sides and a River on one side. The settlement in this village is located at a distance of about 1.5 km from the Coast and protected by the boulders dump (fishery harbor) along the Coastline. The village Manavely is located at a distance of 500 m from the Coast at an elevation of about 3 m above MSL.The backwater (parallel to the Coast) with boulder pitched bund of 3 m height perpendicular to the coast and the coconut plantations present in the dunes parallel to the Coast restricted the tsunami waves entering the village.

The structural damages in the villages have been evaluated and plotted in the respective places of the village map which has been already georeferenced. The house-to-house geographical positioning has been done using handheld GPS. When such damages were brought under isopach lines, they are not in straight line. It clearly reflects the tsunami tongue though advanced in a village or coast, the damage may not be uniform in view of the differences prevailed in the relief of the surroundings in which the structure is placed, the nature of construction, the angle by which structure faced the advancing tsunami waves, etc. The nature of damages in Gundu Uppalavadi village of Cuddalore taluk is shown in Fig.2 as example. In the same way some sample photographs depicting 25 to 50%, 50 to 75% and >75% damages are shown respectively for Ariyagoshti, Villiyanallur and M.G.R.Thittu. Similarly, an effort has been made to prepare the damage isopachs to all the 30 villages.

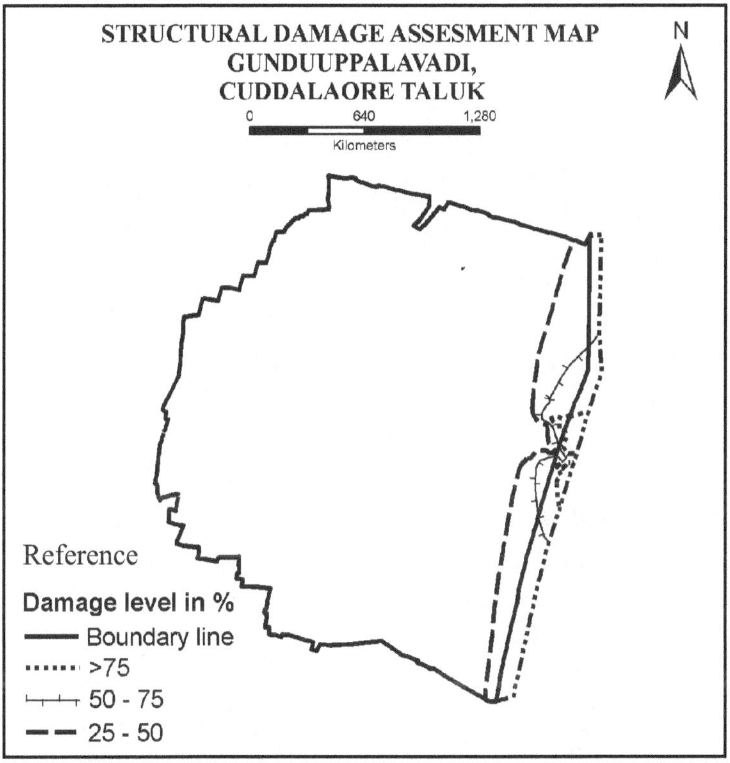

Figure 2: Structural Damage Assessment Map

GEOMORPHOLOGICAL CHANGES

Coastal geomorphology is an important field of study in understanding the impact of tsunami. The coastal landforms are the first features which meet the burnt of fire against tsunami. They record better the nature of tsunami than anything else.

The present study area comprises three Taluks namely Pondicherry in the north, Cuddalore in the center and Chidambaram in the south. These three taluks are displaying unique features of landforms distinctly different to each other. The coastal plain is becoming gradually wider from north to south. While Pondicherry is dominated by sand cliff and terraces, Cuddalore is unique by having parabolic dunes and Chidambaram is different by having series of Beach Ridges with Mangroves and also dominated by paleolagoonal plains. Though, these regions are adjacent to each other, they are unique in having specific features attached to each one of them.

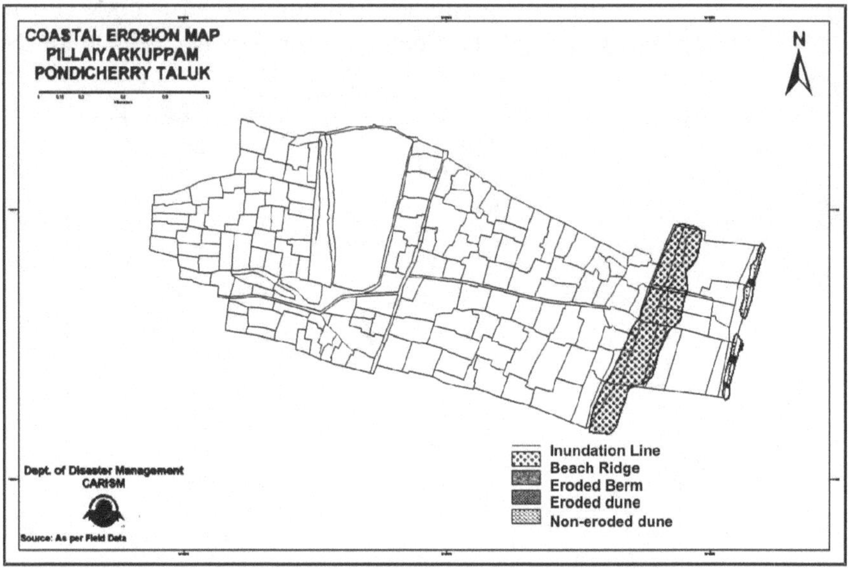

Figure 3: Coastal Erosion Map

Chidambaram Taluk

During pre-tsunami, Chidambaram Taluk maintained a set of landforms which are shown in Fig.4. There are eight series of parallely arranged beach ridges. This coast is found to have two major river courses of Vellar and Coleroon having number of abandoned channels on either sides. Coleroon is manifested with point bars. The alluvial plains in the west are separated from the rest of the Coastal Zone by means of well built cheniers. The Cheniers are noticed to engulf the palaeo-lagoonal plains mainly made up of silty sediments. The Coleroon river and Vellar river in addition to abandoned river channels display a series of natural levee. The rivers in this region register the

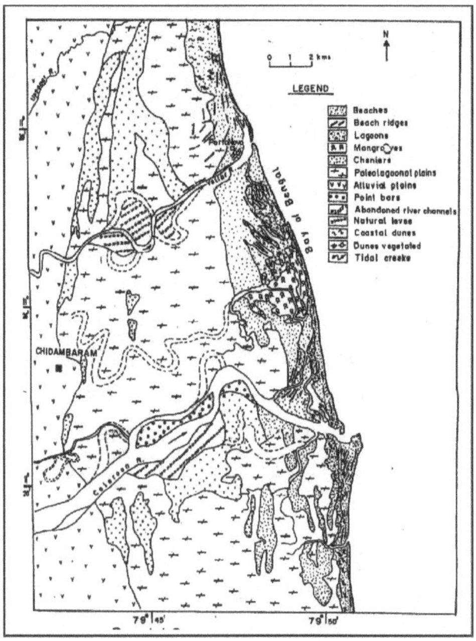

Figure 4: Geomorphology map of Chidambaram region

shifting of the courses by the presence of Palaeochannels on either sides. The Palaeochannels present in line with mangroves is expected to be the possible palaeo course of the Coleroon and hence Pichavaram Mangroves should be the river mouth of the earlier Coleroon, so this area has a series of coastal dunes. The western extreme dunes are invariably seen to have been vegetated, around the Mangroves on either sides. There are lagoons on both northern and southern sides spreading across nearly 1.5 to 3 km distance.

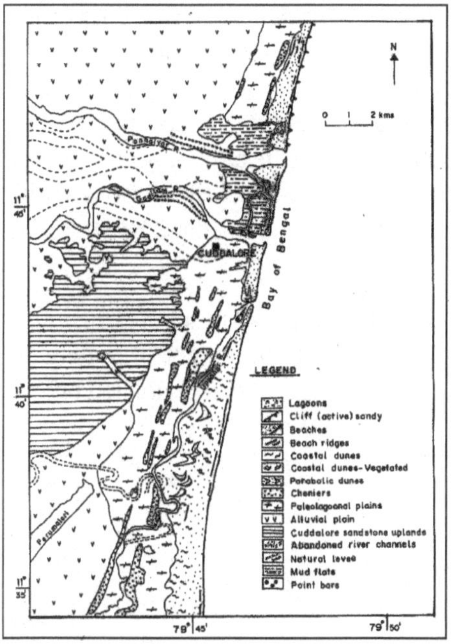

Figure 5: Geomorphology map of Cuddalore region

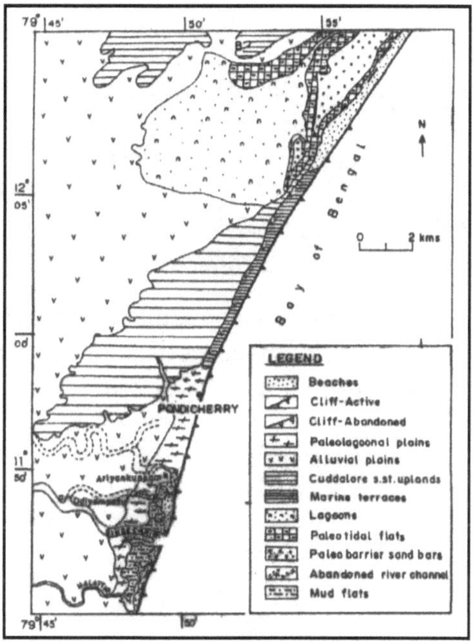

Figure 6: Geomorphology map of Pondicherry region

Cuddalore Taluk

During pre-tsunami, Cuddalore Taluk is noticed to have a set of coastal landforms shown in Fig.5. Unlike Chidambaram, here beach ridges are limited to a maximum of three rows amongst which, the second and third row are seen in the northern extreme. The middle portion of the region is occupied by Cuddalore sandstone. Surrounding the Cuddalore sandstones, alluvial plains are present. These Alluvial plains record the presence of palaeochannels of Ponnaiyar and Gadilam Rivers in the north and Uppanar river in the south. There are parallely arranged cheniers surrounded by palaeo-lagoonal plains. Uppanar forms the boundary line of beach and palaeo-lagoonal plains. In the southern portion it is manifested with parabolic dunes and vegetated dunes whereas the northern portion is free of dunes. The Uppanar river shows the presence of abandoned channels. In the case of,

Gadilam and Ponnaiyar rivers, there are presence of natural levee in the south to the former and in the north to the later. Instead of dunes, the coastal zones at the mouth of rivers of Gadilam and Ponnaiyar give way to the formation of mud flats. The Gadilam river and Uppanar river mouth points out the presence of point bars.

Pondicherry Taluk

During pre-tsunami, the Pondicherry coast (Fig.6) is found to have manifested with marine terraces in the central portion in the place of beach. The western portion of the region is fully occupied by the coastal alluvial plains. Cuddalore sandstone uplands separate the marine terraces and alluvial plains. In the northern part, presence of lagoons surrounded by palaeo tidal flats are seen. There is a palaeo barrier sand bars in the northern extreme. It is seen behind the beach. In the case of south because of the presence of Gingeear river and Malattar river another set of landforms with Palaeochannels, abandoned river channels and mud flats are seen. The shoreline shows the existence of active sandy cliffs.

During post-tsunami, the region under study has undergone severe erosion in the beach, berm, river mouth and dunes (Photos 4 to 7).The extent of damage is displayed in the Table-1. The eroded sands are found to have been deposited behind

Photo 4: Kodiyampalam-Dune

Photo 5: Kodiyampalam-Beach Ridge

Photo 6: M.G.R.Thittu-Dune Erosion

Photo 7: Pillumedu-Berm Erosion

the dunes in the backshore upto a maximum distance of about 700 m. Wherever the inlets, backwater and rivers are there, the sandy sediments are being pushed into much deeper even upto 2.5 km, as in the case of Coleroon. Again the back wash along the course of such inlets has enabled the spread of sandy sediments to even upto 200 m width from the respective banks in the crop fields. The river mouths are seen to have been disfigured. In the case of Vellar river, the mouth has been completely opened by removing the partial sand bars already existing.

LAND USE CHANGES

The pre-tsunami (IRS P6 LISS IV MONO + LISS III Merged data - Fig.7A) and post-tsunami (QUICKBIRD PAN 31 December 2004 – Fig.7B) satellite data were used for studying the land use changes. The pre-tsunami land use and post-tsunami land use, prepared for Gundu Uppalavadi village are shown in Fig.8A and 8B. The damages assessment data for crops and aquaculture farms are shown in Table-2.

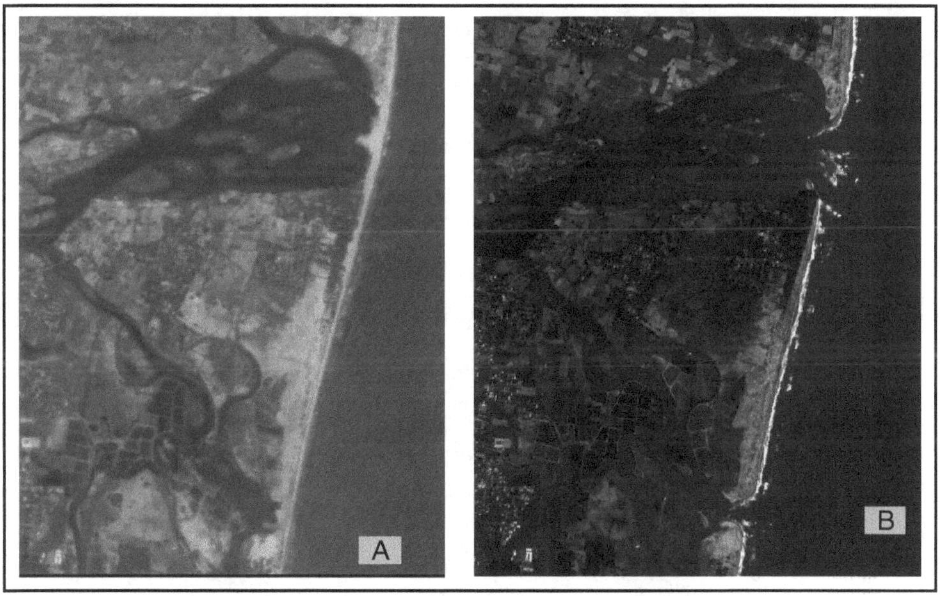

Figure 7: Pre-tsunami (A), Post-tsunami (B), Gundu Uppalavadi,Cuddalore Taluk

HYDROGEOLOGICAL CHANGES

233 water samples from the tsunami affected region (99 from Cuddalore, 101 from Chidambaram and 82 from Pondicherry Taluks) were collected from the field and analyzed for quality parameters with the help of Water Analyzer kit, Flame Photometer, Spectrophotometer, Atomic Absorption Spectrophotmeter and also by Titration. The parameters like pH, sodium, potassium, calcium, hardness,

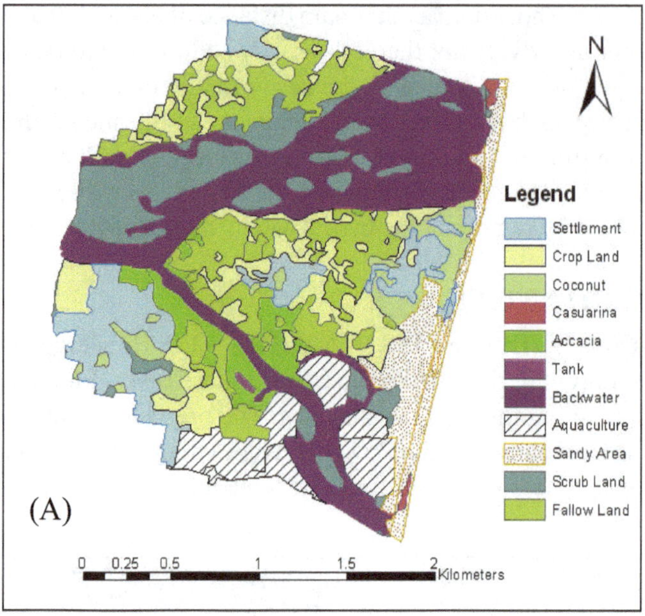

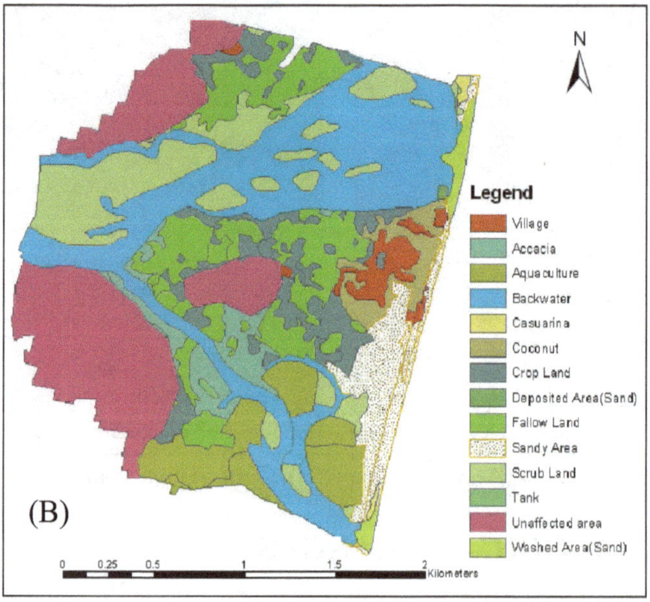

Figure 8: Pre-tsunami (A), Post-tsunami (B), Gundu Uppalavadi, Cuddalore Taluk

magnesium, chloride, aluminium, sulphate and fluoride are found to be beyond the permissible limit prescribed for the drinking purposes in all the taluks. However, Chidambaram and Pondicherry taluks have registered electrical conductivity, dissolved oxygen and alkalinity to higher levels beyond the tolerance limits. The lack of impact in these three parameters in Cuddalore is worth to be noted.

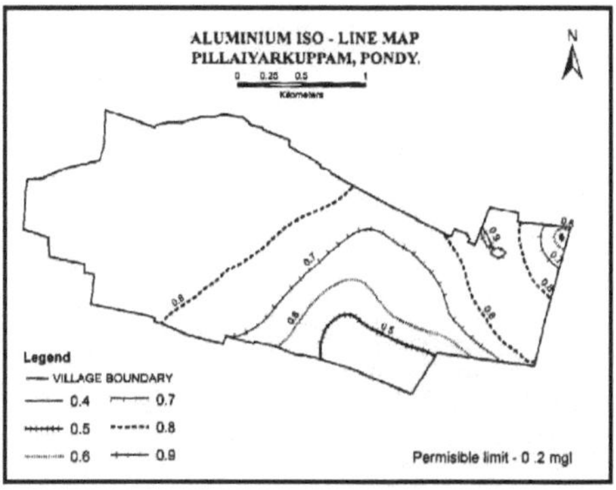

Figure 9a: Aluminium ISO-Line map

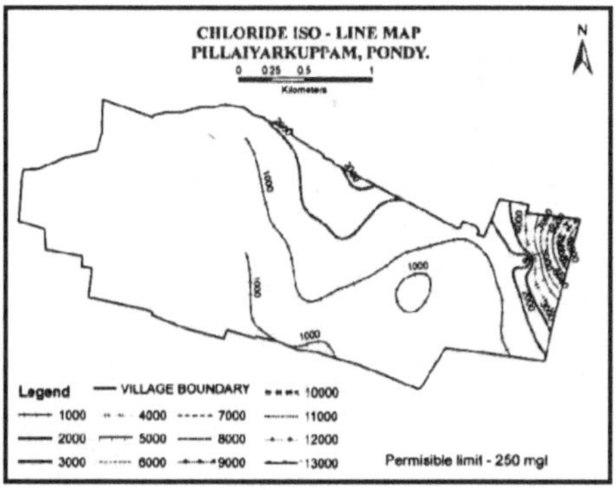

Figure 9b: Cholride ISO-Line map

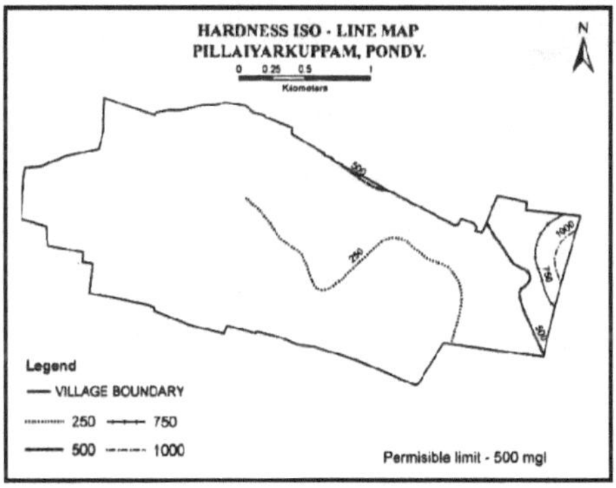

Figure 9c: Hardness ISO-Line map

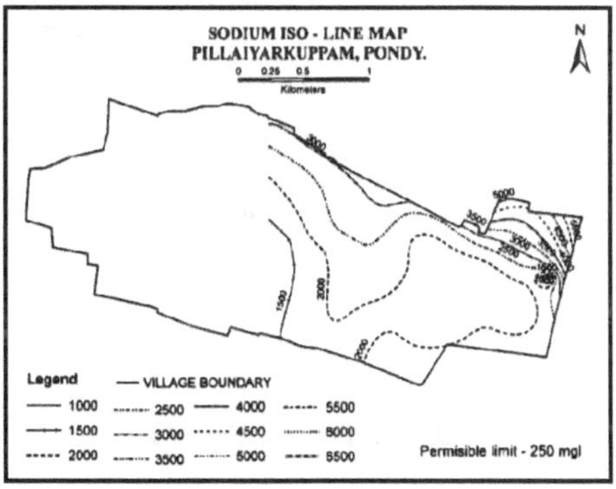

Figure 9d: Sodium ISO-Line map

The samples analysed for Pillaiyarkuppam village are shown Fig.9. The distribution of aluminium (Fig.9a), chloride (Fig.9b), hardness (Fig.9c) and sodium (Fig.9d) display that the isopach lines are not in a rhythmic trend from parameter to parameter. Though, all these parameters shown above are beyond the permissible limits, the extent of the impact varies from parameter to parameter. Similar effort has been made for all the 30 villages and for all the parameters. Detailed analysis may throw light upon the source of the respective parameters and the modality by which the ground water deterioration has taken place.

SOIL SAMPLE ANALYSIS

The soil samples (Numbers 170) were collected from the field and analyzed for quality parameters with the help of Water Analyzer Kit, and Flame Photometer. The parameters like pH, Electrical Conductivity, Alkalinity, Sodium, Calcium and Potassium are found to be within the permissible limits in all the three taluks. The analysed values are shown in Table-3.

Risk Map

An attempt has been made to prepare risk map through the above discussed various analysis. The risk map so prepared for the Ariyagoshti village is shown in Fig.10 as example.

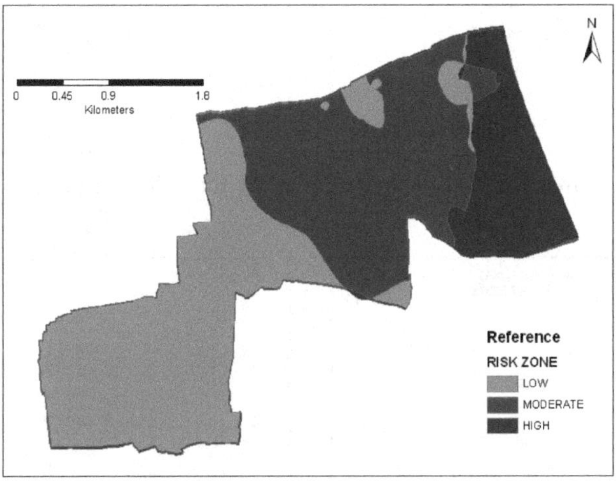

Figure 10: Risk map of Ariyagoshti Village

ACKNOWLEDGEMENT

The authors are thankful to Prof. R. Sethuraman, Vice-Chancellor and his administrative staff members for their unlimited support in completing the studies. The authors acknowledge The Secretary, Department of Science and Technology, New Delhi, Brigadier Dr.R.Sivakumar, Adviser, NRDMS Division and Dr.Bhoop Singh, for the financial support and their co-operation extended throughout this project. The colleagues in Department of Disaster Management and Centre for Advanced Research in Indian System of Medicine (CARISM), SASTRA Deemed University, Thanjavur are also thanked for their excellent support and co-operation in the field as well as in the laboratory. The Chief Wildlife Warden has granted permission to conduct the survey in the Pitchavaram prohibited areas which deserves special acknowledgement from the authors.

Table 1: The Extent of Erosion along the Coast from Pondicherry to Coleroon Mouth

Village Name	Beach		Pre-tsunami Dune			Post-tsunami Dune			Redistribu-tion of eroded sand		Berm erosion		
	L	W	L	W	H	L	W	H	L	W	L	W	D
CUDDALORE													
Devanampattinam	2400	40	450	30	3	70	30	1	100	250	300	40	1
Akkaraikori	2500	75	350	25	3	55	25	2	20	90	0	0	0
Singarathoppu	300	80	200	60	2	17	0-22	1	10	12	0	0	0
Sonankuppam	1700	60	550	65	4	60	30	1	100	330	0	0	0
Kudikadu	1800	40	1250	30	4	140	30	2	15	200	200	25	1
Pachchayankuppam	600	30	400	25	3	40	20	2	40	300	15	30	2
Thiyagavalli	6150	50	5000	110	5	425	50	2	60	1300	410	35	1
Tiruchopuram	1020	40	500	50	3	435	40	1	60	620	0	0	0
Kayalpattu	2400	50	1700	40	3	300	25	1	60	900	280	50	1
Andarmullipallam	1120	30	450	40	1	163	0-30	1	75	850	730	30	1
Gundu Uppalavadi	2750	70	1800	100	3	250	30	3	40	650	175	5	1
Suba Uppalavadi	900	50	650	25	3	110	0-15	3	50	325	175	30	1
Madalapattu	400	50	300	100	4	50	0-75	4	40	250	175	30	2
PONDICHERRY													
Pillaiyarkuppam	900	45	750	75	4	40	75	3	375	25	8	20	1
Pooranankuppam	2600	20	2000	150	4	70	75	3	100	400	0	0	0
Kirumampakkam	1525	100	1200	75	3	102		2	120	650	0	0	0
Ariyankuppam	1600	80	200	50	2	20	35	2	30	75	30	10	1
Thengaithittu	850	100	75	40	2	55	30	1	20	85	0	0	0
Manavely	1450	120	900	40	3	140	30	3	60	650	0	0	0
Kalapet	1525	50	0	0	0	0	0	0	20	560	295	25	1
Manapattu	1650	35	700	75	3	205	45	2	60	475	0	0	0
Pillaichavadi	2725	50	0	0	0	0	0	0	20	900	420	30	1
Pondicherry Town	5000	0	0	0	0		0	0	0	0	0	0	0
CHIDAMBARAM													
Ariyakoshti	700	75	675	50	2	310	50	1	80	475	0	0	0
Kotthattai	3500	70	200	20	1	30	15	1	100	30	0	0	0
Parangipettai	1600	60	300	25	2	145	15	1	60	80	20	8	1
Silambimangalam	1280	100	0	0	0	0	0	0	25	450	90	30	1
Thandavarayacholan pettai	660	25	300	50	2	60	30	2	20	375	425	20	3
Tirukalipalai	Nil	0	0	0	0	0	0	0	0	0	0	0	0
Villiyanallur	800	50	500	25	2	70	25	1	35	360	0	0	0
Killai	4000	75	1950	50	3	575	50	3	80	1250	420	60	2
Pichchavaram	Nil	0	0	0	0	0	0	0	0	0	0	0	0
Periapattu	1600	75	0	0	0	0	0	0	25	375	425	15	1
Kodiyampalayam	1500	75	1175	100	6	835	50	1	700	40	800	50	1

Table 2: The Damages of Crops and Aquaculture Farms

Sl. No.	Name of the Village	Crops in Hectares	Aquaculture Farms in Hectares
CHIDAMBARAM TALUK			
1.	Kill Triukalipalai	167.33	90.55
2.	Ariyagosthi	119.92	-
3.	Thandavarayacholaganpatti	160.75	8.22
4.	Pitchavaram	101.11	-
5.	Killai	75.31	5.31
6.	Periyapattu	55.61	-
7.	Parangipettai	40.73	-
8.	Villiyanallur	34.31	-
9.	Silambimangalam	25.36	-
10.	Kodiyampalayam	12.21	-
11.	Kottattai	7.21	-
PONDICHERRY TALUK			
1.	Kirumampakkam	155.31	-
2.	Puranakuppam	175.21	-
3.	Pillayarkuppam	185.66	-
4.	Manaveli	125.56	-
5.	Pillaichavadi	95.01	-
6.	Manapattu	75.75	-
7.	Ariyankuppam	45.56	-
8.	Pondicherry Town	40.44	-
9.	Thengathittu	35.65	-
10.	Kalapet	30.44	-
CUDDALORE TALUK			
1.	Thiyagavalli	182.40	-
2.	Cuddalore Municipality	62.04	11.71
3.	Kayalpattu	78.05	-
4.	Gundu Uppalavadi	61.04	-
5.	Tiruchchopuram	27.72	-
6.	Kudikadu	25.11	-
7.	Andarmullippalam	28.60	-
8.	Suba Uppalavadi	19.84	-
9.	Madalapattu	8.66	

Table 3: Soil Analysis

CUDDALORE TALUK	pH	EC	Alkalinity	Na	K	Ca
Kudikadu	7.50	0.91	105.00	368.00	42.9	268
Kudikadu	6.89	0.04	35.00	13.80	3.9	12
Pachayankuppam	7.38	0.51	30.00	71.30	7.8	38
Pachayankuppam	6.91	0.03	20.0	11.50	3.9	22.0
Thiyagavalli	7.07	0.99	25.0	184.00	7.8	24.0
Thiyagavalli	6.76	0.03	25.0	11.50	3.9	10.0
Thiyagavalli	6.95	0.22	20.0	50.60	3.9	18.0
Thiyagavalli	6.65	0.03	25.0	13.80	3.9	14.0
Thiyagavalli	7.59	0.07	45.0	16.10	3.9	16.0
Thiyagavalli	6.85	0.02	30.0	11.50	3.9	8.0
Thiyagavalli	5.32	0.03	30.0	9.20	7.8	12.0
Thiruchopuram	7.23	0.65	45.0	299.00	7.8	26.0
Thiruchopuram	6.87	0.03	30.0	11.50	3.9	8.0
Thiruchopuram	7.08	0.02	35.0	11.50	7.8	16.0
Kayalpattu	7.43	0.50	40.0	9.20	7.8	22.0
Kayalpattu	6.92	0.02	50.0	11.50	3.9	8.0
Kayalpattu	8.51	0.88	40.0	391.00	31.2	240.0
Kayalpattu	7.51	0.03	60.0	11.50	3.9	12.0
Andarmullipallam	7.82	0.74	25.0	230.00	7.8	30.0
Andarmullipallam	7.33	0.05	60.0	13.80	3.9	12.0
Andarmullipallam	8.15	1.13	45.0	230.00	7.8	26.0
Andarmullipallam	7.47	0.07	40.0	11.50	3.9	6.0
Gundu Uppalavadi	7.88	1.03	25.0	184.00	7.8	212.0
Gundu Uppalavadi	6.52	0.05	45.0	13.80	7.8	24.0
Gundu Uppalavadi	7.86	0.17	65.0	34.50	117.0	202.0
Suba Uppalavadi	6.52	0.05	60.0	13.80	3.9	38.0
Suba Uppalavadi	7.59	1.03	45.0	230.00	7.8	38.0
Suba Uppalavadi	6.65	0.07	55.0	13.80	23.4	34.0
Suba Uppalavadi	7.23	1.75	55.0	644.00	50.7	50.0
Suba Uppalavadi	7.13	0.10	45.0	16.10	50.7	54.0
Suba Uppalavadi	6.89	1.38	80.0	598.00	156.0	136.0
Madalapattu	7.51	0.14	60.0	18.40	46.8	48.0
Madalapattu	6.91	0.06	45.0	13.80	50.7	44.0
Madalapattu	6.32	0.03	30.0	13.80	7.8	28.0
Madalapattu	6.86	0.04	25.0	11.50	3.9	46.0
Madalapattu	6.80	0.04	35.0	11.50	3.9	20.0
Madalapattu	8.28	0.09	50.0	27.60	11.7	40.0
PONDY						
Pillaiyarkuppam	6.33	2.27	70.0	345.00	7.8	52.0
Pillaiyarkuppam	6.50	0.17	55.0	4.60	3.9	48.0
Pillaiyarkuppam	6.47	0.13	45.0	4.60	3.9	54.0
Pillaiyarkuppam	6.46	0.14	50.0	9.20	0.0	110.0
Pillaiyarkuppam	6.30	0.09	40.0	11.50	3.9	50.0

	pH	EC	Alkalinity	Na	K	Ca
Pillaiyarkuppam	6.11	0.07	45.0	9.20	7.8	20.0
Pillaiyarkuppam	6.50	0.06	45.0	16.10	3.9	74.0
Pillaiyarkuppam	6.33	0.34	60.0	0.00	15.6	64.0
Poornamkuppam	6.53	0.52	30.0	9.20	0.0	14.0
Poornamkuppam	6.74	3.20	50.0	529.00	35.1	48.0
Poornamkuppam	6.46	0.09	50.0	9.20	3.9	12.0
Poornamkuppam	6.69	0.08	45.0	6.90	7.8	28.0
Poornamkuppam	6.65	0.10	55.0	16.10	31.2	40.0
Poornamkuppam	6.64	5.11	95.0	1092.50	46.8	90.0
Poornamkuppam	6.58	1.03	50.0	276.00	3.9	28.0
Poornamkuppam	6.62	1.18	55.0	345.00	7.8	72.0
Poornamkuppam	6.69	0.10	50.0	13.80	3.9	14.0
Kirumampakkam	6.87	0.08	50.0	11.50	0.0	40.0
Kirumampakkam	6.41	1.81	45.0	460.00	3.9	30.0
Kirumampakkam	6.82	0.16	45.0	9.20	3.9	22.0
Kirumampakkam	6.75	2.45	80.0	506.00	42.9	72.0
Kirumampakkam	6.55	0.19	55.0	18.40	50.7	54.0
Manavely	6.64	2.67	50.0	299.00	3.9	26.0
Manavely	7.00	0.13	45.0	0.00	0.0	14.0
Manavely	6.64	0.11	45.0	0.00	0.0	8.0
Manavely	6.79	0.07	45.0	18.40	0.0	12.0
Manavely	6.43	0.06	45.0	13.80	0.0	34.0
Manavely	6.84	0.88	40.0	230.00	3.9	28.0
Manavely	6.55	0.09	40.0	0.00	0.0	12.0
Manavely	7.08	0.07	35.0	11.50	0.0	6.0
Manavely	7.03	0.43	50.0	41.40	0.0	34.0
Manavely	6.82	2.65	55.0	345.00	7.8	66.0
Thengathittu	6.33	1.61	45.0	391.00	15.6	26.0
Thengathittu	7.01	0.82	90.0	253.00	78.0	250.0
Thengathittu	6.63	1.78	60.0	391.00	19.5	616.0
Kalapet	6.47	0.89	45.0	161.00	3.9	28.0
Kalapet	6.87	0.12	45.0	13.80	3.9	12.0
Kalapet	6.31	0.07	55.0	16.10	39.0	44.0
Kalapet	6.90	0.08	40.0	16.10	0.0	28.0
Kalapet	6.47	0.30	50.0	36.00	23.4	102.0
Kalapet	6.58	0.10	45.0	16.10	23.4	40.0
Kalapet	5.95	0.19	50.0	6.90	11.7	26.0
Kalapet	6.63	0.07	35.0	9.20	3.9	12.0
Kalapet	6.40	0.87	40.0	368.00	7.8	98.0
Kalapet	6.40	0.08	45.0	11.50	19.5	30.0
Pillaichavady	6.68	0.99	40.0	345.00	3.9	60.0
Pillaichavady	6.73	0.17	55.0	6.90	7.8	52.0
Pillaichavady	6.64	0.19	35.0	6.90	3.9	16.0
Pillaichavady	6.39	0.38	40.0	57.50	7.8	28.0
Pillaichavady	6.54	0.06	50.0	13.80	0.0	4.0
Pillaichavady	6.09	0.65	80.0	41.40	23.4	90.0

	pH	EC	Alkalinity	Na	K	Ca
CHIDAMBARAM						
Kothattai	6.62	2.43	40.0	529.00	7.8	28.0
Kothattai	6.96	0.09	35.0	161.00	13.9	30.0
Kothattai	6.58	0.15	65.0	2.30	3.9	30.0
Kothattai	6.62	0.06	35.0	6.90	0.0	18.0
Kothattai	6.49	0.08	40.0	16.10	3.9	38.0
Kothattai	6.53	0.09	50.0	16.10	7.8	38.0
Ariyagoshti	7.02	0.27	50.0	36.80	7.8	56.0
Ariyagoshti	7.26	0.09	55.0	16.10	3.9	34.0
Ariyagoshti	6.94	0.33	50.0	39.10	7.8	44.0
Ariyagoshti	7.19	0.13	45.0	16.10	7.8	54.0
Ariyagoshti	7.13	0.13	50.0	13.80	7.8	48.0
Ariyagoshti	6.92	0.70	45.0	16.10	0.0	26.0
Ariyagoshti	6.57	0.11	35.0	0.00	0.0	16.0
Ariyagoshti	6.69	0.12	60.0	9.20	3.9	22.0
Ariyagoshti	6.65	1.02	50.0	299.00	3.9	50.0
Ariyagoshti	6.45	0.06	50.0	16.10	0.0	30.0
Ariyagoshti	6.15	3.45	60.0	1265.00	15.6	36.0
Ariyagoshti	6.40	0.22	65.0	11.50	11.7	84.0
Ariyagoshti	6.46	0.17	55.0	0.00	11.7	54.0
Ariyagoshti	6.20	5.64	55.0	920.00	11.7	26.0
Villiyanallur	6.55	2.15	45.0	552.00	7.8	32.0
Villiyanallur	6.82	0.07	50.0	13.80	3.9	32.0
Villiyanallur	6.41	0.11	55.0	16.10	3.9	26.0
Villiyanallur	6.73	0.13	80.0	13.80	15.6	40.0
Silambimangalam	6.33	0.19	55.0	16.10	3.9	10.0
Silambimangalam	6.58	0.08	65.0	13.80	0.0	14.0
Silambimangalam	6.79	0.10	60.0	6.90	3.9	32.0
Silambimangalam	5.97	0.11	60.0	4.60	7.8	50.0
Silambimangalam	6.93	0.09	65.0	16.10	0.0	6.0
Silambimangalam	6.57	1.51	50.0	483.00	3.9	96.0
Silambimangalam	6.49	0.13	65.0	0.00	11.7	52.0
Periyapattu	6.49	0.06	55.0	16.10	0.0	16.0
Periyapattu	6.68	0.09	55.0	16.10	0.0	30.0
Periyapattu	6.72	0.11	55.0	13.80	11.7	12.0
Periyapattu	6.36	0.12	65.0	6.90	0.0	8.0
Parangipettai	6.32	2.18	55.0	483.00	7.8	148.0
Parangipettai	6.03	5.33	65.0	690.00	57.3	742.0
Parangipettai	6.96	0.28	75.0	50.60	39.0	290.0
Parangipettai	6.18	0.18	70.0	20.70	19.5	180.0
Killai	3.33	0.24	40.0	13.80	3.9	14.0
Killai	6.57	0.26	55.0	18.40	15.6	46.0
Killai	3.30	0.74	35.0	92.00	3.9	34.0
Killai	4.03	0.89	45.0	207.00	3.9	32.0
Killai	3.78	0.18	55.0	23.00	3.9	26.0

	pH	EC	Alkalinity	Na	K	Ca
Killai	6.76	0.39	70.0	62.10	3.9	32.0
Killai	6.19	1.23	50.0	345.00	7.8	30.0
Killai	6.46	0.13	55.0	4.60	3.9	32.0
Killai	6.28	0.10	45.0	4.60	3.9	34.0
Killai	6.63	3.55	65.0	782.00	11.7	40.0
Killai	4.85	2.30	65.0	621.00	15.6	34.0
Killai	6.75	2.79	60.0	736.00	27.3	44.0
Killai	7.29	0.24	75.0	23.00	19.5	68.0
Killai	7.14	0.28	60.0	23.00	7.8	30.0
Killai	6.28	0.18	55.0	4.60	11.7	28.0
Killai	6.73	2.09	55.0	437.00	11.7	58.0
Killai	7.05	0.19	60.0	9.20	3.9	44.0
Killai	6.94	0.20	60.0	11.50	3.9	26.0
Killai	7.22	0.28	65.0	18.40	11.8	40.0
Killai	6.75	2.12	55.0	368.00	3.9	44.0
Killai	7.19	0.17	55.0	4.60	0.0	30.0
Killai	4.96	0.14	45.0	2.30	0.0	70.0
Killai	6.40	4.04	50.0	644.00	11.7	30.0
Killai	6.26	2.68	60.0	690.00	11.7	92.0
Killai	6.53	2.05	45.0	460.00	7.8	58.0
Killai	6.03	0.23	45.0	9.20	0.0	24.0
Killai	4.42	2.41	40.0	437.00	7.8	20.0
Killai	6.53	1.48	45.0	322.00	3.9	44.0
Killai	6.33	3.68	50.0	460.00	7.8	66.0
Killai	6.58	3.71	60.0	92.00	3.9	36.0
Killai	6.35	9.08	50.0	1552.50	23.4	92.0
TSpettai	5.59	3.12	45.0	667.00	27.3	36.0
TSpettai	4.35	3.94	55.0	621.00	7.8	90.0
TSpettai	6.65	0.30	45.0	18.40	3.9	44.0
TSpettai	6.97	17.70	75.0	4140.00	230.1	174.0
TSpettai	7.39	0.25	75.0	11.50	19.5	326.0
Kil Tirukalipalai	7.17	0.34	90.0	18.40	15.6	48.0
Kil Tirukalipalai	7.03	8.65	80.0	2070.00	50.7	102.0
Pichavaram	6.72	0.70	85.0	15.60	66.3	76.0
Pichavaram	6.36	6.73	70.0	1495.00	222.3	150.0
Pichavaram	6.48	9.08	60.0	2012.50	35.1	90.0
Pichavaram	6.41	0.13	45.0	2.30	0.0	34.0
Pichavaram	6.30	0.12	45.0	2.30	3.9	28.0

GEOMATICS IN TSUNAMI
S.M. Ramasamy, C.J. Kumanan, Brig. R. Sivakumar and Bhoop Singh (eds.), pp. 107-134
New India Publishing Agency, New Delhi (India)
E-mail : spjain_niph@rediffmail.com Web: www.bookfactoryindia.com

9

GEOMATICS IN TSUNAMI DAMAGES, NAGAPATTINAM COAST, INDIA

SM.Ramasamy, C.J.Kumanan, K.Palanivel, R.Selvakumar, V.Agastheeswaran, D.Sankar, Steny Thomas and J. Saravanavel

Abstract

Geomatics (Remote Sensing and GIS) based tsunami damage assessment was carried out for Nagapattinam coast as a part of All India Coordinated Programme of NRDMS, Department of Science and Technology, New Delhi. The paper discusses the methodology and the inferences made for Nagapattinam region, Sirkali block and north Poigainallur panchayat village, holistically for the former and as case studies for the latter two. The regional studies show that the coastal geomorphology has played a significant role in tutoring the tsunami inundation and the damages with beach, dunes, tidal flats, backwaters and rivers / streams facilitating the inundation of tsunami. The studies in Sirkali block show the similar role of geomorphology, but the damage assessment carried out for Water bodies, soil and ground water show appreciable damage to Aquifer in the form of saltwater mixing / intrusion due to tsunami. Detailed geospatial data bases were generated village wise too. On the basis of the studies, possible remedial strategies were also suggested.

Keywords: Remote Sensing, GIS, Tsunami 2004, Damage Assessment, Tsunami mitigation, Nagapattinam.

INTRODUCTION

The studies carried out by the geoscientists during the past few decades have brought out a hierarchy of information on the plate tectonics including the occurrence of all seven continents together as one piece of land namely “Pangaea”, the fragmentation of this land, the movement of these continents in different directions around 230 million years ago etc. The studies have further revealed that the Indian plate alone has traveled to the tune of 6000 km in 60 million years towards north northeasterly. Because of the movement of Indian plate and its collision with the Eurasian plate, the Indian plate has subducted below, the northern Eurasian plate has resultantly buckled and The Himalayas has phenomenally risen along the margin of Indian and Eurasian plates. This Himalayan subduction zone extends for several thousand kilometers viz: Himalayas in the north, Arakanyoma hill ranges of Myanmar in the northeast, Andaman in the east, Sumatra in the southeast and further on, towards south easterly with reference to present geo-position of India.

Similar plate margins are found in different parts of the globe with convergent and divergent phenomena. The convergent plate margins are similar to Himalayan-Sumatra belt wherein the two plates are colliding with each other causing subduction and overriding of the colliding plates. Where as along divergent plate margins, the plates on either sides are pulling-a-part from each other. The way mountain belts are formed along convergent plate margins/subduction zones, mid oceanic ridges, ocean floor spreading etc., are formed along divergent plate margins. Between these two types of plate margins, the convergent plate margins or the zones of tectonic subduction are more seismogenic because of collision of plates and that too when these occur in ocean floor, these can trigger huge tides called "Tsunami". The Pacific Ocean is the maximum generator of such tsunamis due to the presence of largest length of subduction zones and hence most of the territorial nations of the Pacific Ocean are always under the grip of tsunamis. But in contrast, it has been a very rare phenomenon in Indian Ocean, despite the presence of a very long seismogenic subduction zone along Rangoon – Andaman – Sumatra, which is regularly witnessing earthquakes and tremors of > 6 M. But the scientists and the administrators of many Asian regions, including India, were not concentrating much on tsunami preparedness because of its rarity and insignificant loss.

However, the major plate subduction of the Indian plate below the Burmise micro- plate along Andaman – Sumatra has caused an earthquake of > 9 M (Ritcher scale) which inturn triggered a massive tsunami on 26th December, 2004 causing colossal devastation to man and his properties in most of the coastal nations of south Asia, including the east coast of India. Along the east coast of India, the state of Tamil Nadu and Union Territory of Pondicherry were worst affected followed by Andhra Pradesh and southern part of Kerala, besides Andaman and Nicobar Islands. Subsequent to the tsunami, a number of inferences were made. Kumanan (2005) has brought out information on the global scenario of tsunamis that out of 1226 tsunami events, 1020 occurred in Pacific ocean, 144 in Atlantic ocean and only 62 in Indian ocean. He further observed that the frequency of the tsunami seems to be on the increase and during the last 500 years alone over 1135 tsunamis have been triggered by the earthquakes of magnitude 7 – 8 M. Somasundaram (2005) has cited references from the Tamil literature that a phenomenon called "Kadalkol", that is "swallowing up of land" by the sea, which may be the mega-tsunamis and were known atleast since 3000 years before present (Y.B.P) in the southern part of Indian peninsular. From the vulnerability point of view, Radhakriashna (2005) has attributed the massive damage along Nagapattinam coast of Tamil Nadu to the crossing of Palaghat – Cauvery shear zone, a major tectonic feature. Narayana (2005) made a rapid appraisal in parts of Kerala coast and observed a broad relation between the geomorphology and tsunami inundation. Ramasamy et al (2005) inferred that the coastal geometry in the form of convexities and concavities have played a significant role in tsunami inundation along the east coast of India, with the former diverting and latter convecting the tsunami waves. Ramasamy et al (2006) further have made a regional appraisal along the east coast of Tamil Nadu and inferred that geomorphology has played a major role in absorbing, accommodating and carrying the tsunami surge on to the land. Several observations were also made by many, on the possible impact of tsunami over the natural resources (Ramasamy 2005, Devarajan 2005) and environmental systems (Usha Natesan 2005, Murugesan et al 2005, Kannan et al 2005).

As tsunami is totally a new phenomenon and not much of information is available from generic or from vulnerability points of view, the NRDMS Division of DST has launched an "**All India Coordinated Programme**" for tsunami damage assessment with an aim of evaluating the damages along the east coast of India and Andaman – Nicober islands to create GIS based large scale digital data bases on the same, so that not only the damages are assessed in detail, but also the phenomenon is understood and strategies are evolved for futuristic tsunami mitigation. The authors have taken up this programme for the coastal region of Nagapattinam district and the salient findings are discussed in this paper.

The Nagapattinam coastal segment taken up in the present study is located in the central part of the north-south to north northeast – south southwest trending 1000 km long Tamil Nadu coast. Conspicuously, the studied coastal segment forms straight north – south alignment extending from south of Pichchavaram in the north to Point Calimere in the south (Fig.1). This coastal segment fringes easterly the major Cauvery bird foot shaped arcuate delta. A number of east-west faults are reported in the area (Elliot et al 1998, Ramasamy et al 2006) which are currently occupied by creeks. In addition, there are a number of southerly, southeasterly, easterly and northeasterly flowing drainage systems respectively in the southern Vedaranniyam, central Nagapattinam – Pumpukar and the northern Tirumullaivasal – Pichchavaram areas. However, the Karaikal region is not covered in this present study.

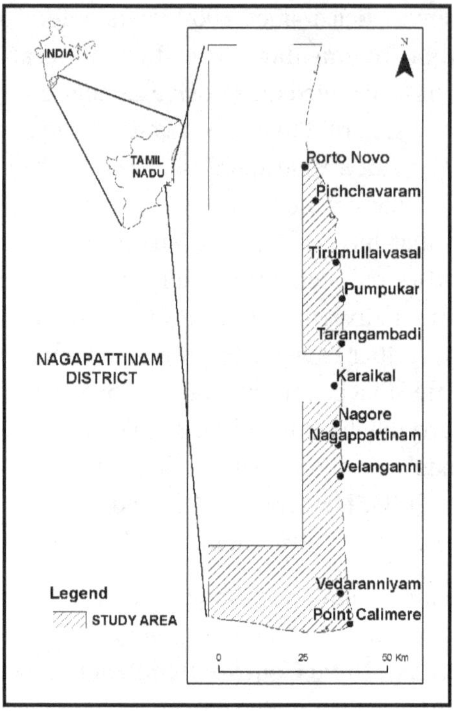

Figure 1: Study Area

METHODOLOGY (Fig.2)

In the present study, tsunami damage assessment was made on three aspects, viz:

- Natural resources
- Physical resources (facilities/ development sector wise)and
- Human resources

In the case of natural resources, the damages caused by the tsunami in geomorphology, land use / land cover, soil, surface water/wet lands and groundwater systems were analysed. For these, the pre-tsunami IRS P6 PAN merged digital data dated 5th March & 21st May 2004 and the post-tsunami ERS data (Earth Resources Satellite of European Space Agency) dated 29th Dec 2004; just 3 days after tsunami, were used. Though, the ERS data had coarser resolution has spectacularly showed tsunami inundated and the receded zones too due to its microwave spectral region. However, for detailed damage assessment in respect of geomorphology, land use/ land cover, soil, water etc., deserving field studies were undertaken to evaluate the type and extent of damages. In addition, E.C (Electrical Conductivity) and pH of soils and soft mud underlying the water bodies and wet lands were also analysed by collecting samples at 10-20 cm depth. In the case of groundwater, data on various hydrogeochemical parameters collected from the archives of State Surface and Groundwater Data Centre, Government of Tamil Nadu, were analysed and from the same EC values were studied in detail, for the years 1992, 1995, 2000 and 2004 (pre - tsunami period) and 2005 (post -tsunami period). The post-tsunami EC measurements were carried out by the investigators from the hand pumps and bore wells. Similarly, the data on the damages of physical resources in different

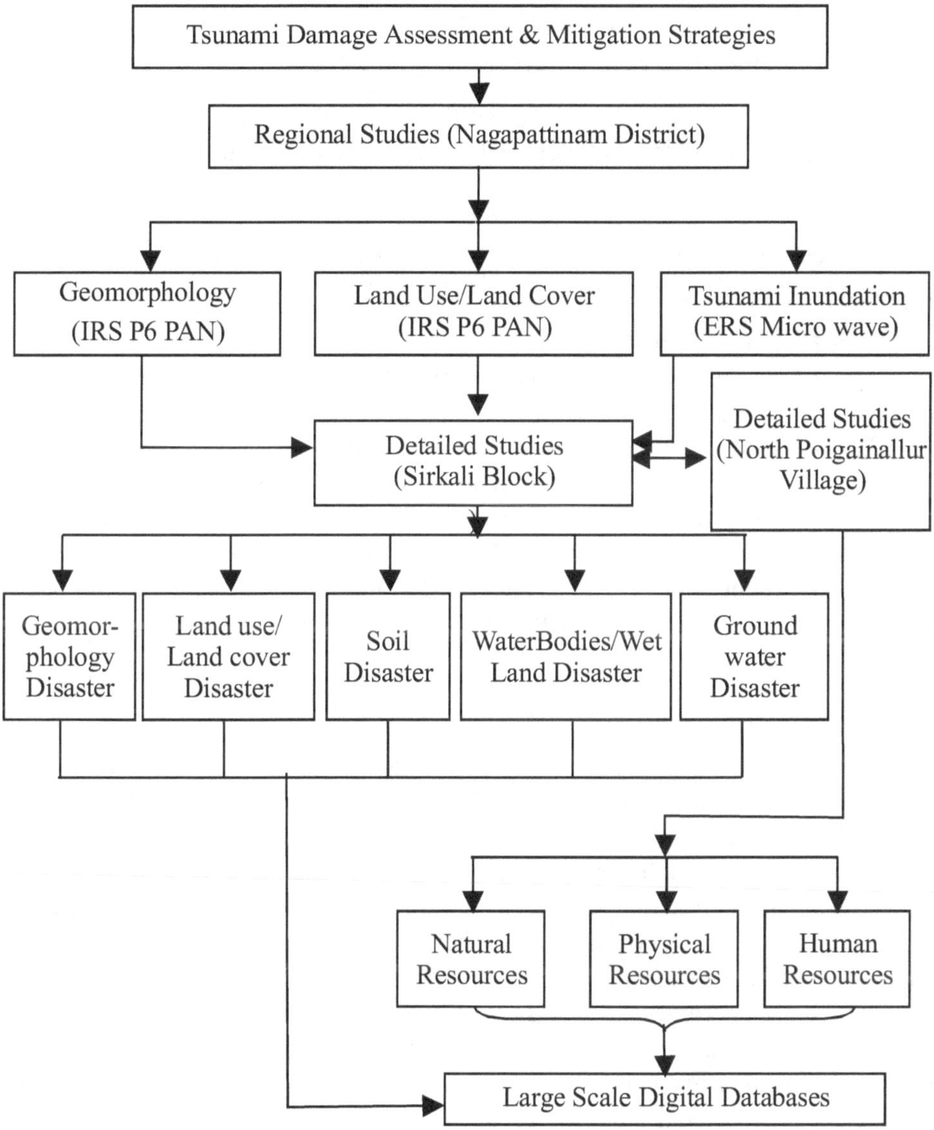

Figure 2: Methodology

developmental sectors namely: education, health, irrigation, transport, communication, agriculture etc., were collected through a specially devised format. The data collected for human resources damages were analysed in terms of death and missing. Finally, geospatial databases were generated on the above damages.

However, this paper deals with the damage assessment carried out

- Regionally for the whole coast of Nagapattinam region (1:50,000 scale)
- Locally for one block, namely Sirkali, on 1:8000 scale (for natural resources damages only - as a case study) and
- Microscopically for one village, namely north Poigainallur of Nagapattinam Block on 1:8000 scale (as case study again)

From these, three tier studies, possible and futuristic mitigative strategies were also evolved.

TSUNAMI DAMAGES – REGIONAL ASSESSMENT – (NAGAPATTINAM DISTRICT)

Tsunami Inundation (Fig.3)

The ERS data was optically zoomed to 1:50,000 scale after proper georeferencing, and the area of tidal inundations were marked for the study area (Fig.3). Such tsunami inundation map was digitized using Arc-GIS and GIS image was created. The Cursory look has indicated that the tsunami waves have inundated up to 500 – 600 meters from the shoreline all along the coast. But the tidal waves have gushed on to the land for several kilometers along the rivers, streams, creeks, palaeo and present backwaters etc. As Nagapattinam region has a large number of such features, tidal inundation has covered fairly a larger breadth all along the coast.

Geomorphology (Fig.4 & 5)

For the regional assessment, the geomorphology and land use / land cover data were used as indices and the detailed geomorphology was interpreted using IRS P6 PAN merged digital data by zooming it to 1:8000 scale and all the geomorphic features were identified. In general, the fluvial, fluvio-marine and marine landforms were observed in the area (Fig.4 & 5). The fluvial landforms were represented by rivers, channel bars, flood plains, meander plains, abandoned channels etc. The fluvio-marine landforms were mostly of deltas,creeks, palaeoswales overprinted by deltaic sediments and sand sheets etc. Whereas the marine landforms interpreted were estuaries, spits, bay mouth bars, present and palaeo backwaters, salt flats, present and palaeo mudflats, thinly developed beaches, palaeo beach ridges etc. Synoptically, the fluvial landforms were predominantly seen in the west, fluvio-marine landforms in the central and the marine landforms all along the coast in the

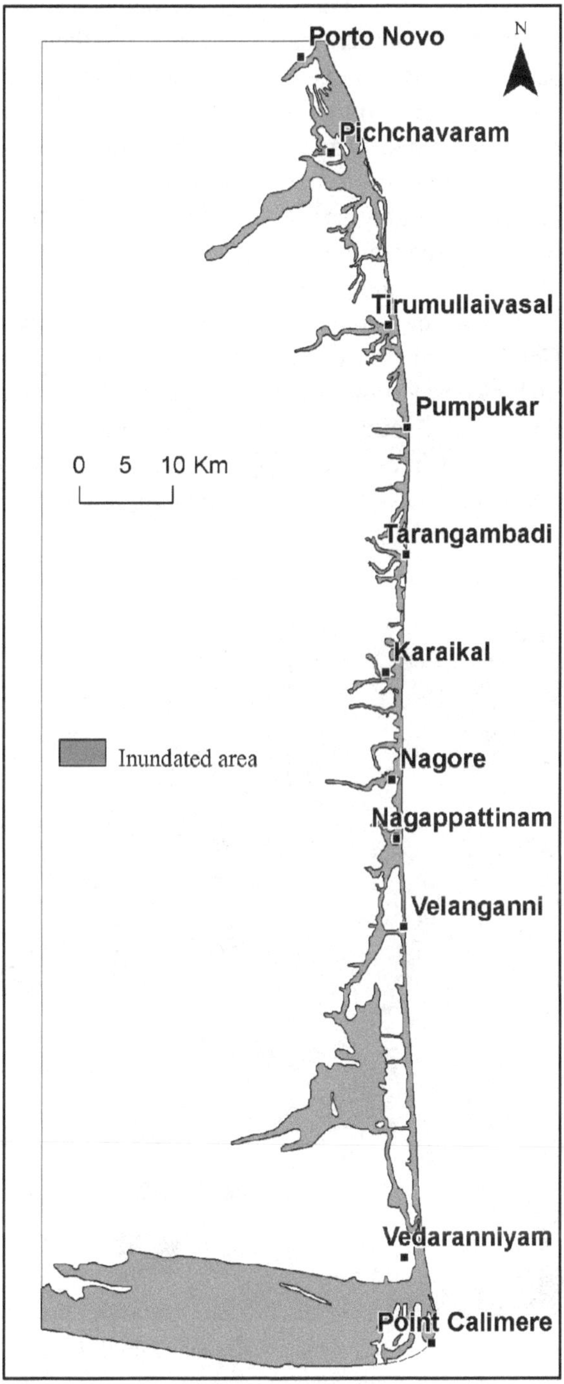

Figure 3: Tsunami Inundation in Parts of Nagapattinam District (Source ERS data)

east; all the three having N-S sub parallelism. As stated earlier, the creeks were found more frequently in the southern part of the area and these creeks were found along the active east-west faults (Elliot et al 1998, Ramasamy et al 2006). Ramasamy et al (2006) have further observed that while the streams and creeks found to the south of Nagapattinam town don't have bay mouth bars, they appear to be more to the north Nagapattinam along the coast. Conspicuously, the palaeo beach ridges as usual were trending as long and linear sand ridges from almost Porto Novo in the north to Point Calimere in the south (Fig.5). Two such broad palaeo beach ridges with intervening swales were interpreted in the study area (18,Fig.5). While these beach ridges and swales were seen to a breadth 3-4 km along the coast in the entire stretch, in Point Calimere area, these were seen as bundles nearly to a breadth at 35-40 kilometers in between Tiruthuraipoondi in the northwest and Point Calimere in the southeast (Fig.4). In the southern most rim, there is a major backwater namely Vedaranniyam - Muthupet backwater (6,Fig.5) with a length of around 55-60 kilometers and a breadth of 12-14 kilometers. These geomorphic features were digitized using Arc-GIS and GIS data base was created.

(CK-Creeks, BW- Backwater, SF-Salt flats, BC-Beach, PRMF-Present mud flats, PLMF, Palaeo mud flats, PSOD-Palaeo swales overprinted by delta, PBR-Palaeo beach ridges)
Figure 4: Satellite Image Showing Geomorphology (Vedaranniyam Region)

The GIS image of tsunami inundation (Fig.3) was superimposed over the geomorphology GIS image (Fig.5) and the estimations were done on the aerial percentage of inundation in each of the above landform categories. Such a GIS Add function analysis has revealed that the marine landforms like estuaries, spits, backwater, tidal flats and creeks and the fluvial landforms like rivers and streams were predominantly inundated to the tune of more than 85% of their aerial extent. Whereas, the landforms like abandoned channels, palaeo swales, palaeo swales overprinted by delta, palaeo beach ridges etc. were inundated to less than 20% of their aerial extent. From this, it can be concluded that regionally, the marine landforms and the rivers of the fluvial landforms have succumbed more to tsunami inundation.

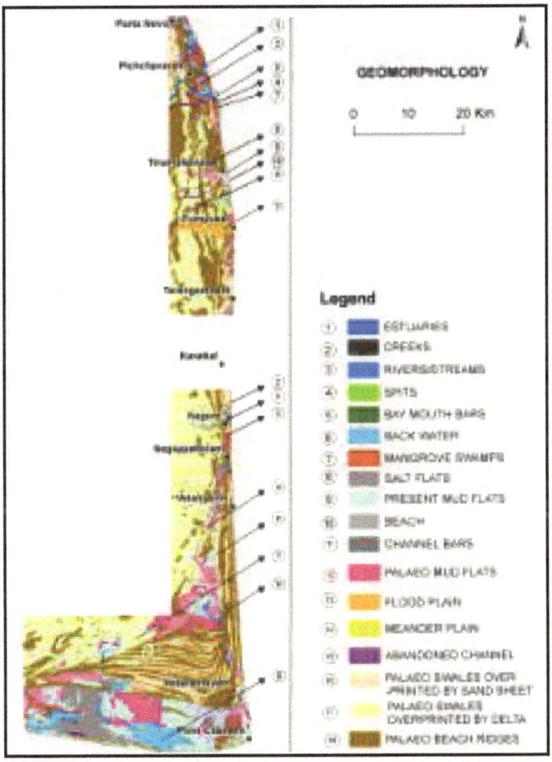

Figure 5: Geomorphology- Nagapattinam Region

In general, most of the beaches have been flattened, spits were destroyed and bay mouth bars were washed off. The beach dunes having an elevation of 3-5 meters prior to tsunami were also flattened. The tsunami waves have dumped the marine sediments brought by them in backwaters, palaeo and present mudflats and along rivers / streams too; however along the drainages not much of sediments were seen during the survey, probably due to their removal during the backwash of the tsunami waves.

Land Use / Land Cover (Fig.6)

The land use / land cover details were similarly interpreted from IRS P6 PAN merged data by enlarging it to 1:8000 scale using ArcGIS and various features were interpreted through on screen digitization (Fig.6). The wet lands namely the creeks, palaeo and present backwaters, palaeo and present mudflats, river beds, meander plains etc., were mostly represented as water logged land, marshy land, salt flats, aquaculture ponds etc., The beaches were interpreted as sandy area. The land use / land cover features like plantations and settlements were mostly seen in the elevated long and linear beach ridges. Whereas, wherever the deltaic sediments found overprinting the palaeo swales, crop lands and fallow lands were seen.

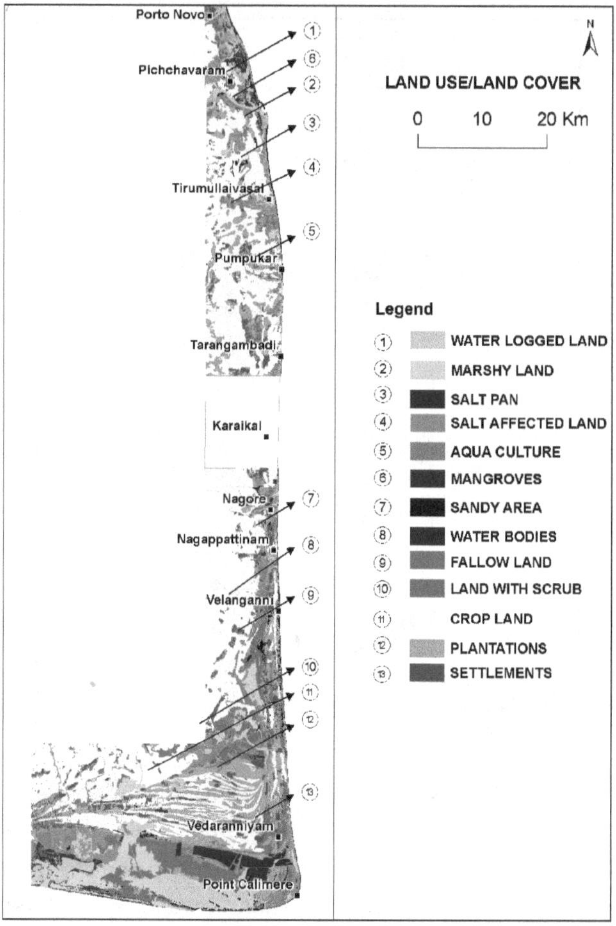

Figure 6: Land Use / Land Cover – Nagapattinam Region

Similarly, the tsunami inundation GIS image (Fig.3) was superimposed over the land use / land cover GIS layer (Fig.6) and area percentage of each of the inundated land use/ land cover category were estimated. From the same, it is inferred that mostly the water logged and marshy lands (wet lands), salt flats and aquaculture zones (present and palaeo mudflats) were inundated to the tune of 75 – 90 %. Thus from the above analysis, it can be said that the land use / land cover not only act as indices in assessing the natural resources damage but also can provide guidelines in carving out futuristic mitigation strategies.

TSUNAMI DAMAGES – BLOCKWISE - DETAILED ASSESSMENT – (SIRKALI BLOCK)

Having built the concept on the inundation pattern and therefrom the interface dynamics between tsunami waves and coastal geomorphology and land use/land cover, through the above regional analysis, detailed block wise studies were carried out to assess the damages on natural resources, as the regional studies convincingly showed that the above two (geomorphology and land use/ land cover) can give actual information on the damages to natural resources. In this block wise study, the assessment done for Sirkali

block is narrated in the paper as a case study. In the blockwise analysis, natural resources damages were assessed in terms of geomorphology, land use/land cover, soil, water bodies / wet lands and groundwater.

Tsunami Inundation (Fig.7)

As done in the case of regional assessment, the area inundated by tsunami was mapped by interpreting the enlarged format of ERS data for Sirkali block and the same is shown in (Fig.7)

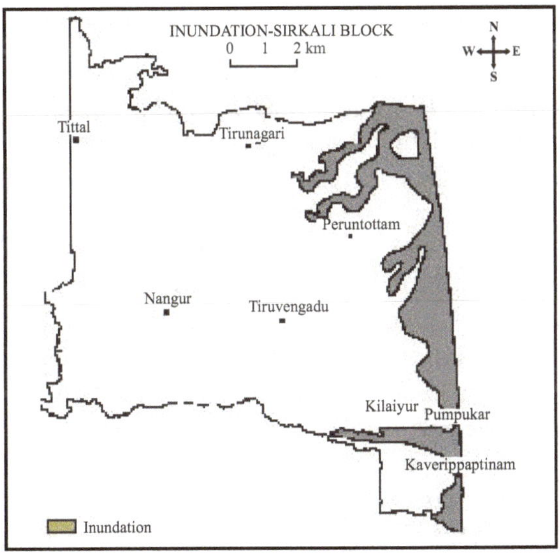

Figure 7: Tsunami Inundation – Sirkali Block (Source ERS data)

Geomorphology (Fig. 8 & 9)

The detailed geomorphology map was prepared similarly by enlarging the IRS P6 PAN merged data to 1:8000 scale and the same is shown in Fig.8. The detailed interpretation has led to the finer classification of the landforms, interpreted regionally, into two or three classes as shown below in Table-1.

Table 1: Regional vs Block level Geomorphology

S. No.	Regional Geomorphology	Block Level Geomorphology
1	Palaeo Swales Over Printed By Delta	(3classes) • Palaeo swales over Printed by delta • Palaeo swales over Printed by delta / Aquaculture • Palaeo swales over Printed by Delta/ water bodies
2	Palaeo beach ridges	(2classes) • Palaeo beach ridges • Relict palaeo beach ridges
3	Beach	(2classes) • beach • beach under nourishment
4	Present mud flat	(3classes) • Present mud flat • Present mud flat /aquaculture • Present mud flat/salt flat

Similar GIS add function analysis was carried out between tsunami inundation (Fig.7) and geomorphology (Fig.8) of Sirkali block. From such an integrated image, the area percentage of inundation for each category of geomorphology was worked out. For example, the total area of creek in Sirkali block was 0.4812 Sq.km (kept as 100 %) and area of creek inundated by tidal waves was 0.3512 Sq.km and thus from the same, the area percentage of inundation (73%) was worked out. Similarly, for each geomorphological unit, the area percentage of inundation was worked out (Table–2). Such an analysis revealed that the features like swales, present mudflats, beaches, palaeo swales overprinted by delta / aquaculture, relict palaeo beach ridges, beach under nourishment, creeks, rivers etc were inundated to more than 60-100 % of their aerial extent. The post-tsunami field survey has indicated that, the tsunami waves have totally flattened and washed away the beach and also even deflected and uprooted the stone pillars of the barbed fences which indicates the ferocity of the tsunami (Fig.9A). In the same way, even very long and elevated beach dunes were mostly washed away and only relict and remnant dunes could be seen in many places (Fig.9B). As the beaches and the coastal dunes were flattened / washed off along the coast of the Sirkali block, in addition to tsunami, this area has now become vulnerable for inundation even during high tides. In some beaches, heavy mineral layering was seen after tsunami (Fig.9C). This warrants detailed studies because, as the tsunami ferociously hit the coast at the speed of about 40 -50 km/hr and engulfed the coastal lands, the erosional and depositional dynamics of the tsunami might have been a sudden and a vibrant phenomenon and hence such fine layering may not be possible; the load casts like features also confirm the micro events of high energy waves in between. However this may need further studies.

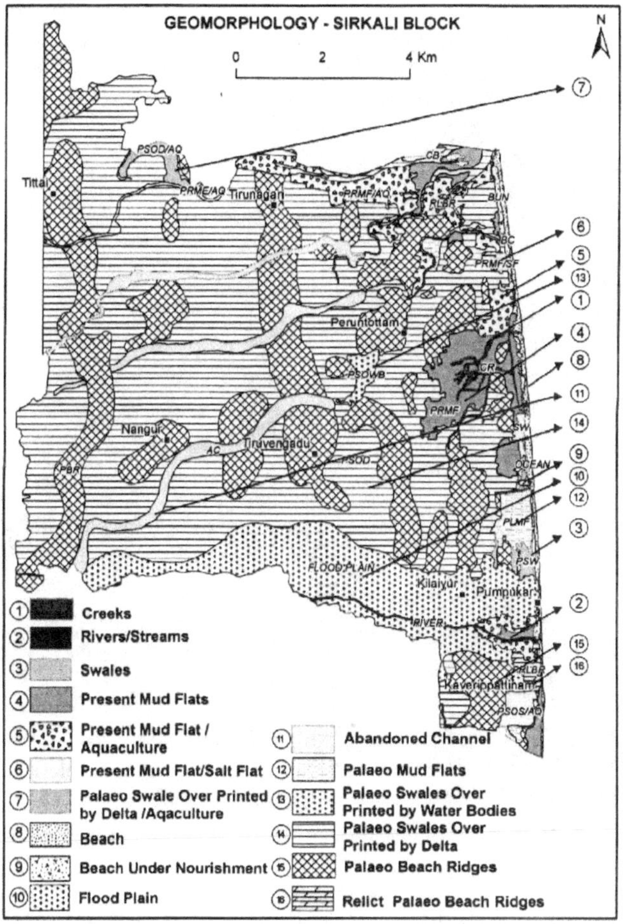

Figure 8: Geomorphology – Sirkali Block

In almost all the river mouths, the bay mouth bars were washed off and in few rivers tsunami cut terraces bearing tsunami dumped garbages were seen even at 2-3 km inside the coast (Fig.9D).

(A) Flattened Beach and Deflected Stone Pillars

(B) Relict Dunes

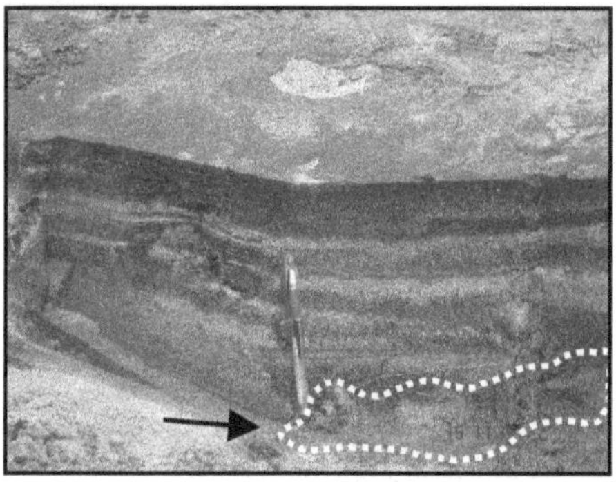

(C) Layering of Heavy Minerals with load casts

(E) Dried up Mudflats

(D) Tsunami Cut Terraces in River Bank

Figure 9: Effects of Tsunami over Geomorphology in Sirkali Block

The EC (Electrical Conductivity) and pH measurements were done in some land- forms using a handheld EC & pH meters which gave the EC values of 100 to 1400 millimhos for the geomorphic features (Table - 2). The lesser EC value of 100 – 1400 millimhos indicates subsequent flushing as the area has witnessed heavy rain and flood after the tsunami. Similarly, the pH measurements carried out in some of the selected landforms showed variance of 6.1 to 7.79 (Table-2). The lower pH in features like present mudflats indicates the acidity due to tsunami inundation. However not much of abnormality were noticed in pH.

Table 2: Geomorphology vs Tsunami (Sirkali Block)

Geomorphology	Total Area (in Sq.km) 100%	Area % of Inundation	Ec (milli mhos)	pH
Creek	0.4812	73		6.78
River	0.2639	60		
Swales	0.1870	100		
Present Mud Flats	4.5355	65		6.1
Present Mud Flats/ Aquaculture	4.6312	51		7.62
Present Mud Flat/ Salt Flat	0.1622	100		
Palaeo Swales Over Printed By Delta/ Aquaculture	0.5406	90		7.18
Beach	1.0957	100	1400	6.98
Beach Under Nourishment	0.5399	96		
Flood Plain	11.9433	11		
Abandoned Channel	4.6046	4		
Palaeo Mud Flat	1.2519	93		
Palaeo Swals	0.3794	79		
Palaeo Swale Over Printed By Delta	54.9165	9		
Palaeo Beach Ridges	31.9026	6	100	7.65
Relict Palaeo Beach Ridges	0.2134	81	100	7.79

Land Use / land cover (Fig.10 & Fig.11)

Similarly, land use / land cover details were interpreted for Sirkali block again by digitally zooming the IRS P6 PAN merged pre-tsunami data (Fig.10). In this case too, finer details were brought out in comparison to regional map. Similarly, area percentage of inundation was worked out for each land use / land cover category which indicated that the water logged area and marshy land (wet lands) were inundated to more than 80 % of their aerial extent. Again, the beaches and beach dune plantations of Sirkali block were inundated to the tune of over 94% of their aerial extent (Table-3).

In many places, the garbages were accumulated in concrete buildings, tall trees and in electrical poles too indicating that the tsunami run-up level might have been more than 35 - 40 feet in the area (Fig.11A).

Table 3: Land Use / Land Cover vs Tsunami (Sirkali Block)

Land Use / Land Cover	Total Area (in Sq.km) 100%	Area % of Inundation	Ec (milli Mhos)	pH
Water Log Ged Land	2.7174	81		
Aquaculture	4.2388	69		7.62
Marshy Land	2.8398	89		
Salt Affected Land	7.1772	39		
Sandy Area	1.2121	94	1400	6.98
Water Bodies	0.2350	5		
Land With Scrub	5.0717	15		7.68
Crop Land	65.6237	6	350	7.79
Plantation	0.4063	94		8.78
Mixed Plantations	23.8181	8	900	7.58
Village	6.1072	4	1900	7.39

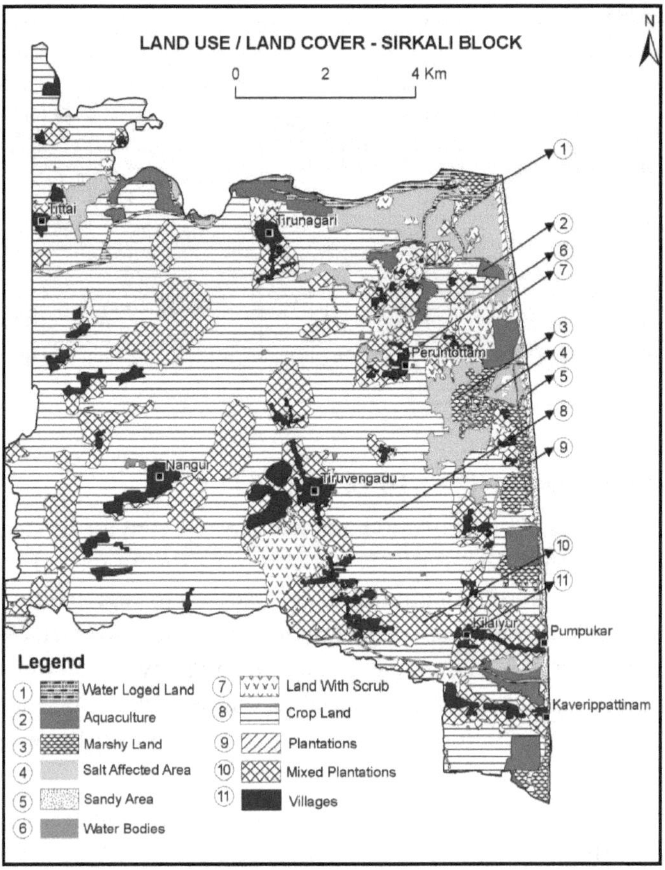

Figure 10: Land use / Land cover- Sirkali Block

Some of the Casuarina plantations developed for nourishing the dunes and beaches were seen uprooted and deflected in most of the areas (Fig.11B). The agricultural fields directly facing the beaches, and facing the beaches through tidal flats, salt pans, aquaculture ponds etc., were blanketed with marine sediments and salt marshes by the tsunami (Fig.11C & 11D).

(A) Garbages In Electrical Pole

(B) Deflected Plants

(C) Salt incursed Agriculture Field

(D) Salt Marshed Agricultural Land

Figure 11: Effects of Tsunami Over Land Use / Land Cover

While EC was observed to be within the permissible limit (Table-3), the pH varied between 6.98 and 8.78; the highest value of 8.78 was witnessed in those tsunami inundated plantation areas (Table-3). Hence such agriculture fields / croplands and plantation lands inundated by the tsunami need to be reclaimed by flushing.

Soil (Fig.12)

Soil map was collected from Tamil Nadu Soil Survey and Land Use Board and GIS data base was created (Fig.12). Broadly, four series of soils have developed in the area namely Valuthalakudi, Thiruvengadu, Adhanur and Kondal. While the Valuthalakudi series was predominantly a clayey soil related to creeks, mudflats, swales, etc., the Thiruvengadu series was seen to be a sandy loamy soil derived from the palaeo beach ridges. Whereas, the Kondal series was a clayey sand and Adhanur series was a sandy clay. Over this GIS image of soil, (Fig.12) the tsunami inundation layer (Fig.7) was superimposed and the area percentage of inundation of each series of soil were worked out and shown in Table-4. On the basis of the tidal inundation, soil samples were collected at 10-20 cm depth both in inundated and un-inundated adjacent areas for comparative analysis of the impact of tsunami over soils. The EC values varied between 500 and 2700 millimhos; falling within permissible limits. However such less EC of 2700-2000 in palaeo tidal flats suggest subsequent flushing by the rains.

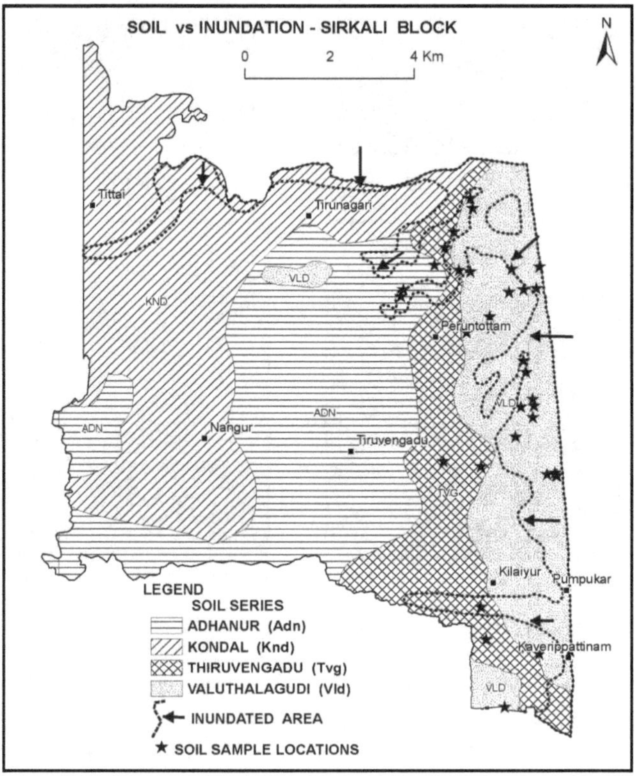

Figure 12: Soil – Sirkali Block

Water Bodies / Wet lands (Fig.13)

Enlarging the toposheets and also enlarging the IRS P6 PAN merged data to 1:8000 scale, all the water bodies / wet lands (rivers/streams, creeks, estuaries, backwaters, surface water bodies etc.,) were interpreted (Fig.13). Over this vector layer, the tsunami inundation layer (Fig.7) was superimposed and the area percentage of inundation of water bodies / wet lands were similarly worked out and shown in Table–5.

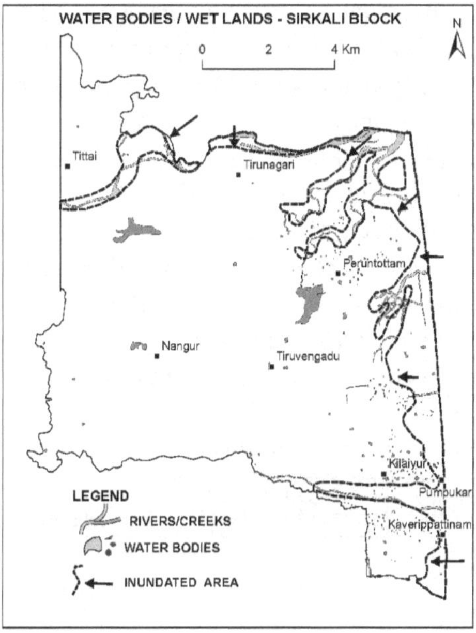

Figure 13: Water Bodies / Wet Lands – Sirkali

Table 4: Soil vs Tsunami (Sirkali Block)

S. No	Soil Series	Total Area (In Sq.km) 100%	Area % of Inundation	EC(milli mhos)	pH
1	VLD	24.501725	47.14	500	6.82
				1400	6.98
				1100	6.93
				500	7.61
				2700	7.05
				1600	7.72
				800	7.71
				1400	8.78
				1100	5.56
				2000	7.68
				600	7.41
				2000	7.62
2	TVG	16.995969	13.69	300	7.52
				400	7.10
				100	7.55
				100	7.22
3	ADN	37.603869	2.72	1900	7.94
4	KND	40.376038	0.42		
*(**VLD**-VALUTHALAGUDI;**TVG**-THIRUVENGADU;**ADN**-ADHANUR;**KND**- KONDAL)*					

This indicates that the tidal inundation was about 51 % in rivers and creeks and 2 % percent in water bodies. Least inundation in water bodies may be due to their far away geo-position from the coast. Moreover, the tsunami water was pumped out from the water bodies; hence not much of impact was noticed. Even the EC was under permissible limit (100 – 1000 millimhos, Table–5). The pH was upto 9.63 in some water bodies which indicates that the beds of these water bodies have become alkaline after tsunami in those cases. In fact, in case of water bodies and creeks, the soft sediment samples were collected at the depth of 10 -20 cm. Hence such water bodies of higher alkalinity may need desiltation during the summer seasons so as to save the water quality in the future. The EC and pH did not show much aberrations in creeks and rivers too even after tsunami inundation.

Table 5: Wetlands vs Tsunami (Sirkali Block)

S.No.	Wet Lands	Total Area (in Sq.km) 100%	Area % of Inundation	EC	pH
1	River/creek	0.018244	51	2000	7.61
				600	7.85
				1200	7.56
				1600	7.72
2	Water Bodies	20.08343	2	100	6.82
				1000	9.63
				600	7.55
				500	7.13
				800	7.71

Groundwater Damage (Fig.14)

For assessing the tsunami damage over the groundwater systems, water samples were collected from over 40 locations, both from tsunami inundated and uninundated areas. As far as the EC values of pre-tsunami period were concerned, the EC data collected from Tamil Nadu Water Supply and Drainage Board were analysed. The EC values were grouped into two categories as less than 3000 millimhos and more than 3000 millimhos (Fig.14A), the former representing the highly desirable limit for drinking purpose and the latter representing the desirable / permissible limit for drinking. GIS images were generated and the same have indicated that only near Peruntottam village there was a small pocket of more than 3000 EC and the other parts of the block mostly fell in good quality zone (Fig.14A). Similarly, the GIS layer was generated for the post-tsunami showing two classes (that is < 3000 and >3000 EC). It was found that after the tsunami, the water quality has deteriorated fairly in a larger area (Fig.14B). The integration of such post-tsunami EC GIS image (Fig.14B) with land use / land cover image (Fig.10) indicated that only in the land use / land cover falling in close proximities to rivers and creeks, the quality got deteriorated. This deterioration has reached the level of 13,000 to 14,000 millimhos in some pockets which need to be desalinated by artificial recharge as the area is getting good monsoon rainfall.

Phenomena of Saltwater Intrusion / Mixing (Fig.15)

The soft Holocene sediments comprising clay, silt and sand and their rhythmites form the aquifers in Sirkali block so the groundwater is under perched and water table conditions. The EC data collected from the State Surface and Groundwater data centre, Government of Tamil Nadu, Chennai for 1992, 2000 and 2004 (pre-tsunami) and 2005 January (post-tsunami) were studied in detail and the boundary lines were drawn independently for each year between <3000 and >3000 (millimhos) EC zones (Fig.15). From the same it is inferred that in between 1992 and 2004 the saltwater front seems to have been receding towards the coast. Where as, in 2005, after the tsunami, the saltwater front has substantially intruded into the land. This was inferred in creeks, streams dominated areas.

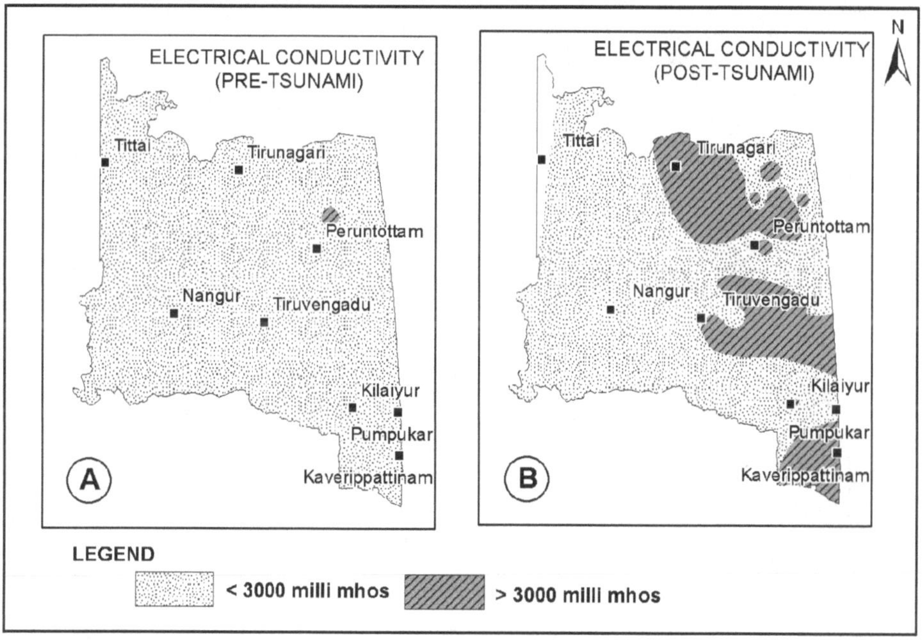

Figure 14: Pre and Post EC – Sirkali Block

TSUNAMI DAMAGES – DETAILED ASSESSMENT – VILLAGE WISE (NORTH POIGAINALLUR) (Fig.16 & Fig.17)

Subsequently, village wise damage assessment was done on natural, physical and human resources using geomatics technology. For example, the studies carried out in north Poigainallur village of Nagapattinam block is discussed here as example. In this, the damages on natural resources were assessed following the methodology adopted for Sirkali block that is on geomorphology, land use / land cover, soil, surface and groundwater resources and the area percentage of damages were worked out and the same is shown pictorially (Fig.16). By and large, the pattern of damage seems to fall in conformity with the block wise interpretation carried out for Sirkali block. As far as this village is concerned, except the area falling in palaeo beach ridges which was slightly elevated, the remaining entire part of the village is a low lying area and so the natural resources have suffered a greater inundation and the same is also reflected in soil and groundwater resources.

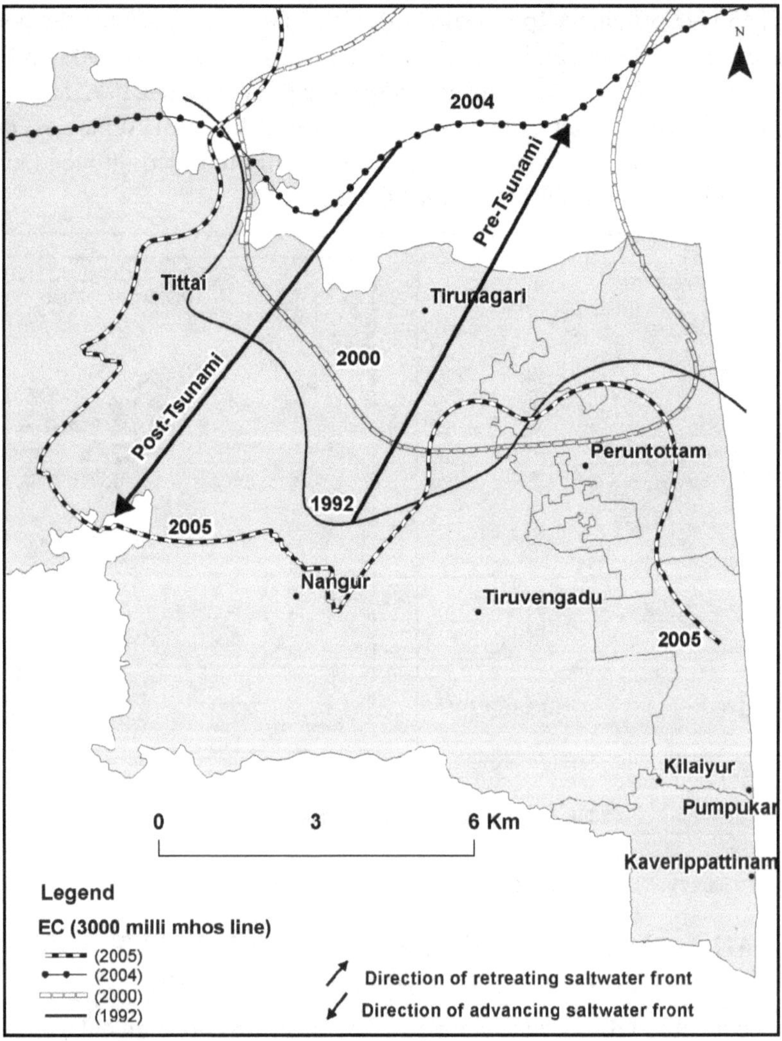

Figure 15: Saltwater Intrusion / Mixing - Sirkali Block

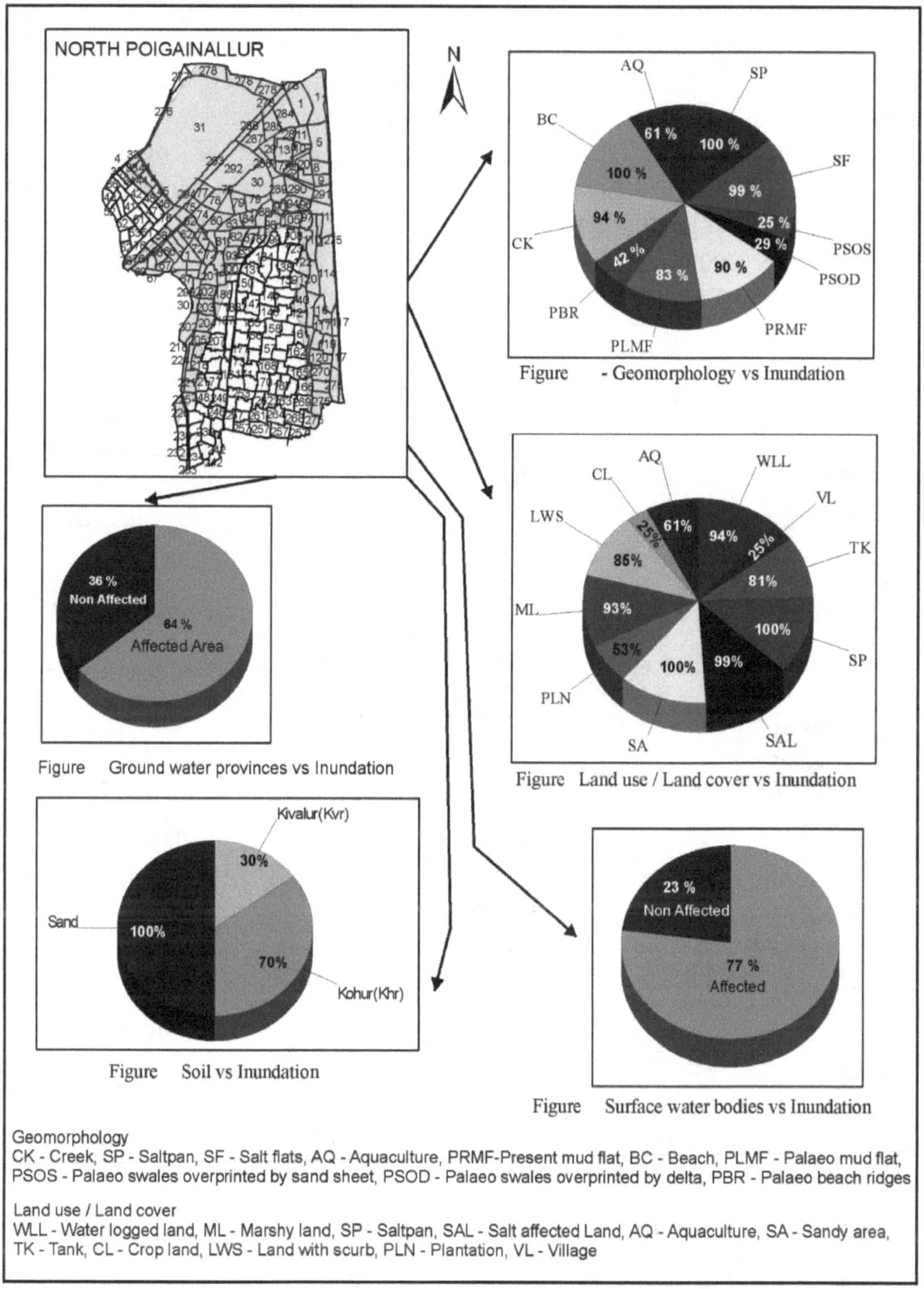

Figure 16: Natural resources – North Poigainallur Village

Similarly, the physical resources damages were analysed for various sectors viz: education, health, irrigation, transport, communication, electricity, public facility, agriculture, property loss etc. In education sector, school buildings were damaged to 34 %, health centers to 47%, irrigation to 20%, transport to 55%, communication to 62 % etc. which have been shown pictorially in Fig.17. In education and health sectors, damages were assessed in terms of buildings. Similarly, transport were assessed in terms of length of roads, communication in terms of length of electric lines and related material destruction, public facility in terms buildings, agriculture in terms of tsunami inundation over the agricultural fields etc. However, as far as the human loss is concerned (death and missing) the total casualties were 15 % of the overall Nagapattinam district.

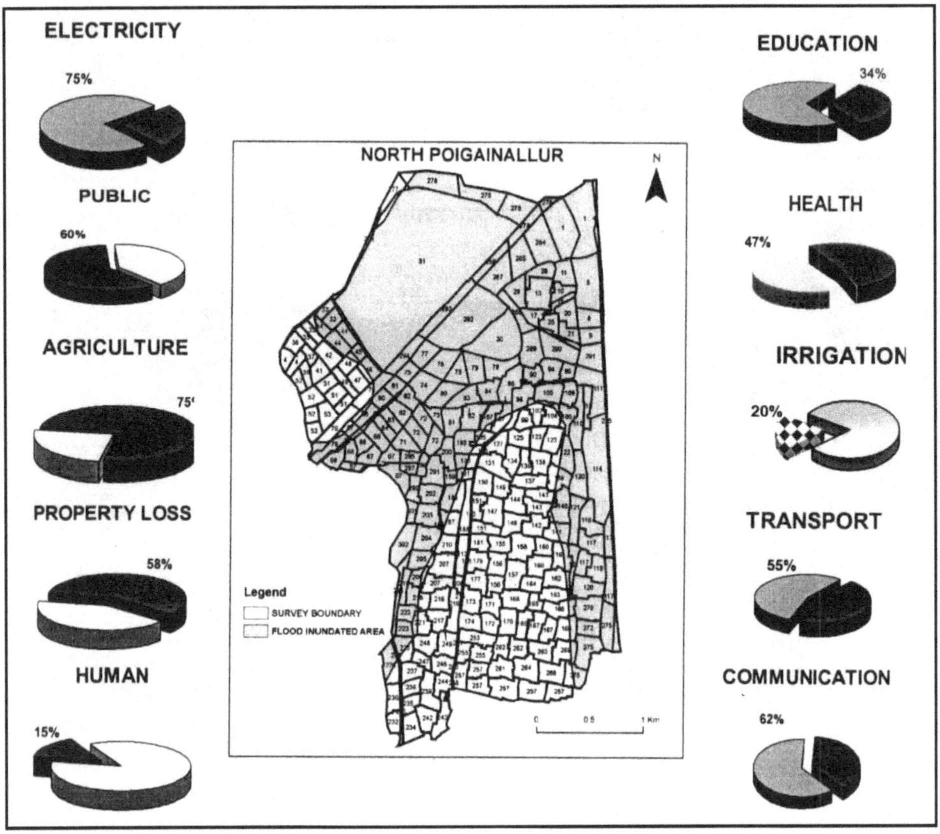

Figure 17: Physical & Human resources – North Poigainallur Village

TSUNAMI VULNERABILITY

The analysis between the tsunami inundation and the geomorphology has indicated that the geomorphology has predominantly controlled the tsunami inundation in Sirkali block and also in Nagapattinam region on the whole. Hence, on the basis of the aerial % of inundation, the geomorphic features were classified into most vulnerable, moderately vulnerable and least vulnerable category (Fig.18).

Mostly, the marine, fluvio-marine and also fluvial landforms especially the rivers/ streams were falling under mostly vulnerable category (1a – 1i,Fig.18). Where as the marine landforms namely palaeo mudflat (2c,Fig.18) and fluvial landforms like flood plain (2a,Fig.18) and abandoned channel (2b,Fig.18) fell under moderately vulnerable category and few landforms belonging to both marine and fluvio-marine landforms were falling under least vulnerable category (Fig.18) as shown in Table-6.

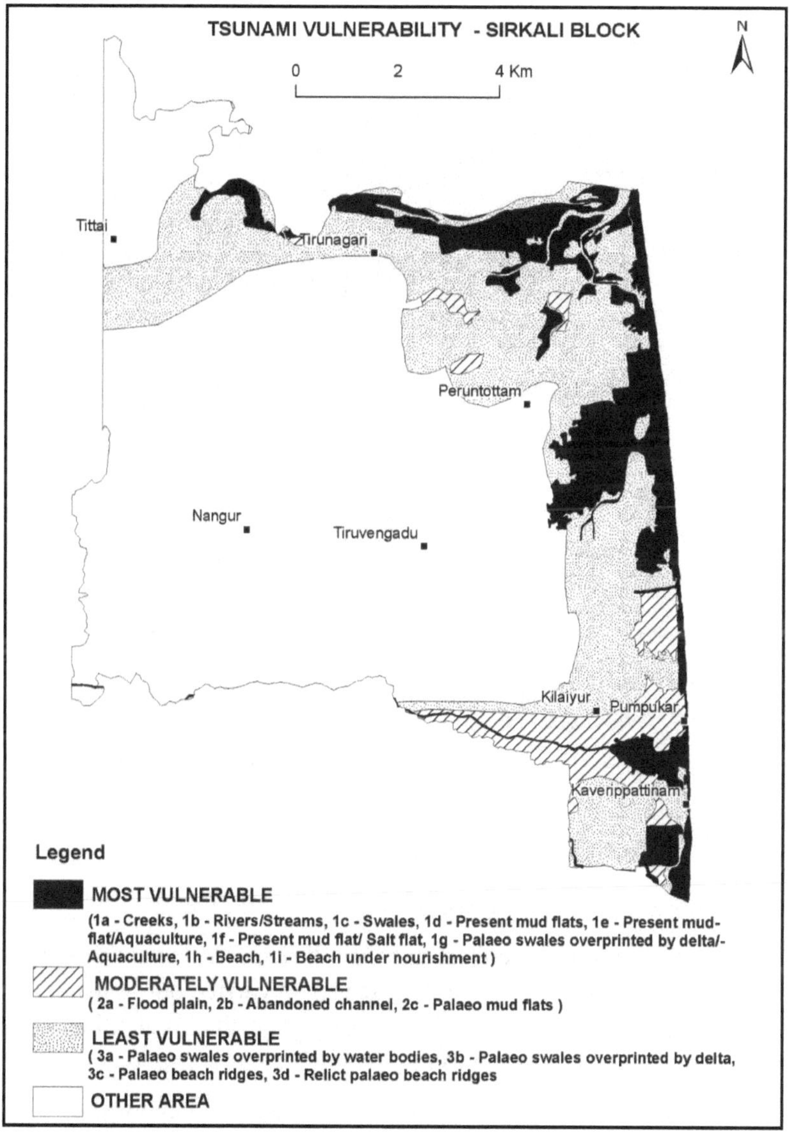

Figure 18: Tsunami Vulnerability – Sirkali Block

On the basis of the vulnerability, the following remedial measures are suggested

Table 6: Tsunami Vulnerability (Sirkali Block)

Category	Features	Remedial Measures
Most vulnerable	Creeks, Rivers / Streams and Swales	• Mouths to be kept open • Stony e mbankments in vulnerable places along the banks of rivers / streams • Mangroves in suitable places
	Present mud flats, Present mud flat / aquaculture, present mud flat / salt flat and palaeo swales overprinted by delta / aquaculture	• If they are directly facing the shore, then stony embankments in between the shore and these features.
	Beach	• Stony embankments in between beach and shore and afforestation in beach
	Beach under nourishment	• Intensive afforestation
Moderately vulnerable	Flood plain and Abandoned channels	• Stony embankment and corridor forestry in between these units and the parent rivers
	Palaeo mud flats	• Afforestation in peripheral areas, bird sanctuaries,etc.

DISCUSSION AND CONCLUSIONS

The present geomatics based (satellite VNIR data, microwave remote sensing, digital cartography, GIS etc.,) damage assessment carried out holistically for the Nagapattinam region, in detail for Sirkali block and north Poigainallur village have brought out the following information.

- The tidal inundation was mostly controlled by the geomorphology and the low lying marine landforms namely beach, beach dunes, wet lands and fluvial landforms like rivers and streams have been extensively inundated by the tsunami waves. In addition, some of the settlements and the irrigated fields which are directly facing the beach or facing the beach through tidal flats, saltpans and aquaculture zones were severely inundated. Accordingly, suitable remedial measures were suggested (Table-6). As the beaches seem to have acted as absorbers of tsunami waves (Ramasamy et al 2006) beaches need to promoted by constructing jetties in suitable locations and nourished through afforestation. The coastal dunes which normally act as barriers need to be stablised by mulching. As rivers / streams have acted as pathways for tidal inundation, their paths are to be kept clean and on the either banks of the streams stony embankments will have to made in vulnerable zones. Some of the villages which have stony embankments (Kannaki statue, Pumpukar) were saved from tsunami attack. However, detailed remedial measures are shown in Table–6.

- The soil did not suffer much to tsunami except alkalinity increase in some areas. But since this area has already witnessed heavy rainfall this would have been flushed out as also seen from the EC values (Table–4).
- The soft sediments collected from water bodies indicated some alkalinity increase in some parts of Sirkali block. These water bodies can be desilted during summer seasons.
- The groundwater quality has deteriorated in Sirkali block wherever the aquifers are in close proximity to rivers/ streams / creeks (Fig.14)
- Again interestingly, in between 1992 and 2004 the saltwater front seems to have receded towards the coast in Sirkali block. But after the tsunami, the saltwater front has advanced towards the land indicating saltwater intrusion (Fig.15).

As far as the village wise studies are concerned, the studies carried out in north Poigainallur showed similar trend with Sirkali block. However, the future detailed analyses are in progress and on completion of the same detailed vulnerability map is to be prepared. Anyway, as far as quality improvement of the groundwater is concerned artificial recharge might probably push the saltwater front towards the sea, in shallow aquifers which provide drinking water supply to the coastal dwellers.

ACKNOWLEDGEMENT

The authors gratefully acknowledge NRDMS division, DST for granting the study and Brig. R.Sivakumar, Head, NRDMS division and Dr. Bhoop Singh, Director, NRDMS are acknowledged for their personal involvement.

REFERENCES

Devarajan, M. (2005) Impact of Tsunami (2004) on Coastal Aquifer Systems-Certain Observations From Tamil Nadu, India. Tsunami: The Indian Context, (eds: SM.Ramasamy and C.J. Kumanan), Allied Publishers Pvt. Ltd., New Delhi, pp. 213-219.

Elliot, C.I., Joyce, E.B., Bishop, I., Ramasamy, SM., & Kumanan, C.J. (1998) Lineaments of Southeastern India and their Relationship to Drainage Pattern and Groundwater Flow. Proceeding Volume on International Association of Hydrogeologists, University of Melbourne, Feb. 8-13, pp 131-135.

Kannan, L., Thangaradjou, T. and Duraisamy, A. (2005) Effects of Tsunami on the Coastal and Marine Biodiversity of India. Tsunami: The Indian Context, (eds: SM.Ramasamy and C.J. Kumanan), Allied Publishers Pvt. Ltd., New Delhi, pp. 261-266

Kumanan, C.J. (2005) Global scenario of Tsunamis. Tsunami: The Indian Context, (eds: SM.Ramasamy and C.J. Kumanan), Allied Publishers Pvt. Ltd., New Delhi, pp. 15-25

Murugesan, A.G., Rajakumari, C. and Sukumaran, N. (2005) Environmental Hazards in Tsunami Hit Coastal areas of Tamil Nadu. Tsunami: The Indian Context, (eds: SM.Ramasamy and C.J. Kumanan), Allied Publishers Pvt. Ltd., New Delhi, pp. 251-260

Narayana, A.C., Tatavarti. R. and Mudrika Shakdwipe (2005) Tsunami of 26 December 2004: Observations on Kerala coast. Journal of Geological Society of India, v.65(3), pp.239 – 246.

Radhakrishna, B.P. (2005). O, horrible! most horrible! – Devastating Tsunami Strikes Coastline of India on 26th December 2004. Journal of Geological Society of India, v.65(2), pp 129 – 134.

Ramasamy, SM (2005) Tsunami in the Indian Context – Certain Issues and Stratigies. Tsunami: The Indian Context, (eds: SM.Ramasamy and C.J. Kumanan), Allied Publishers Pvt. Ltd., New Delhi, pp. 1-14.

Ramasamy, SM, Kumanan,C.J., Selvakumar,R., Saravanavel, J., Anandan, C. and Mahalingam, S. (2005) Tsunami Vulnerability of Tamil Nadu Coast, India – Geological Perspective. Tsunami: The Indian Context, (eds: SM.Ramasamy and C.J. Kumanan), Allied Publishers Pvt. Ltd., New Delhi, pp. 155-162

Ramasamy, SM, Kumanan,C.J., Saravanavel, J. and Selvakumar,R. (2006). Geosystem Responses To December 26 (2004) Tsunami and Mitigation Strategies For Cuddalore – Nagapattinam Coast, Tamil Nadu, India. Journal of Geological Society of India (in press).

Somasundaram, V.N. (2005) Past Marine Incursions: Certain Evidences From Tamil Literature. Tsunami: The Indian Context, (eds: SM.Ramasamy and C.J. Kumanan), Allied Publishers Pvt. Ltd., New Delhi, pp. 27-31.

Usha Natesan (2005) Environmental Damages of Tsunami. Tsunami: The Indian Context, (eds: SM.Ramasamy and C.J. Kumanan), Allied Publishers Pvt. Ltd., New Delhi, pp. 239-249.

GEOMATICS IN TSUNAMI
S.M. Ramasamy, C.J. Kumanan, Brig. R. Sivakumar and Bhoop Singh (eds.), pp. 135-146
New India Publishing Agency, New Delhi (India)
E-mail : spjain_niph@rediffmail.com Web: www.bookfactoryindia.com

10

GEOSPATIAL ASSESSMENT OF TSUNAMI (2004) DAMAGES ALONG THE ROCKY COAST OF KANYAKUMARI, INDIA

N. Chandrasekar , J. Loveson Immanuel, M. Rajamanickam,
J. Dajkumar Sahayam, D. Satheesh Herbert Singh and G.V. Rajamanickam

Abstract

The damages caused by the tsunami surge of 26, December 2004 could not be recuperated. There is no second opinion for the above statement as attested by the consequences of tsunami surge on the coastal landforms, human lives, properties, biological assets, etc., The attempt made to evaluate the damages caused by the tsunami surge along the study area divulges various factors regarding the response of the coastal environment to this weird phenomenon. Kanyakumari district witnessed the tidal inundation of 50 – 450 m, the least along Colachel and the maximum in Kanyakumari and Muttom coasts. Run-up elevations were seen even up to 6 m in Manakudi and Colachel coasts. In major parts, the beaches of the study area had endured lofty erosion of beach sediments. High occurrence of Casualty was reported in the Coasts of Colachel, Kotilpadu, Manakudi and Azhikkal. Similarly, loss of man-made structures were also observed to be high along the Coast of Manakudi, Colachel, Kotilpadu, Azhikkal and Muttom. Costly articles like boats, fishing nets, etc., were also damaged in Kanyakumari, Pallam, Muttom, Kadiapatanam, Kesavanputhenthurai and Colachel. The quality of water has also been found to be affected with high level of pH and TDS in the coasts of Manakudi and Colachel. Such Colossal damages in the area is attributed to its geo-position exactly at the receiving point of the diverted tsunami surge by Sri Lanka. Bathymetry also seems to have played a crucial role in determining the intensity and angle of the seismic wave propagation in the area.

Keywords: Tsunami, India, Damages in Kanyakumari, Evaluation, Map, GIS.

INTRODUCTION

Everything – the Nature and the Man were performing as usual in the study area until the arrival of the 'uninvited guest', the tsunami. Ever since the incident, the customary functioning of the humans and the natural phenomena were disturbed to a considerable extent. Though the natural phenomena have convalesced, the normal functioning of the human beings might probably take a long time. The earthquake measuring >9 M occurred at Sumatra has triggered these tsunami waves and these waves have almost disastered all the territorial nations of South East Asia and parts of Africa too. This study is an attempt to evaluate the total damages including the natural, physical, chemical, biological, environmental and economical aspects along the Kanyakumari coast using geospatial technology. The damage assessment thus done could be taken as a vital input for carving out mitigation strategies. It also provides some critical evidences for the preferential damages in certain segments of the coast, that is, the response of the beaches to such an unexpected hazard, the factors provided resistance to the beaches to face the tsunami and the negative factors contributed to the vulnerability of the coast. The paper discusses certain salient aspects on these.

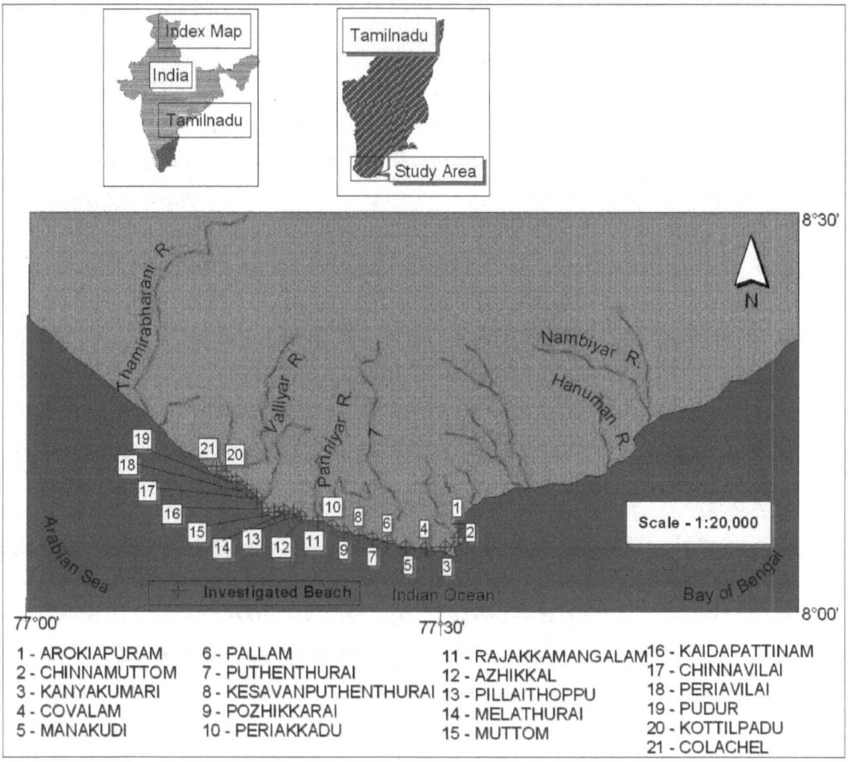

Figure 1: Location Map of the Study Area

REGIONAL SETTING

The study area (Fig.1) lies between Latitudes 8^0 04' 30" and 8^0 12'00" N and Longitudes $77^0$15'45" and $77^0$33'58"E. It falls on the southern coast of Kanyakumari District, Tamilnadu, India. The study area is bounded by Indian Ocean in the South, Arabian Sea in the West and Bay of Bengal in the East but the main part of the coast faces the Arabian Sea with mountains and undulating valleys in the Northern side. The study area is manifested with marine terraces, sand dunes, beachridges, estuaries, mangroves, floodplains, beaches, peneplains, uplands, sea cliff, etc., Apart from the perennial river Thamirabharani, streams like Palaiyar, Panniyar, Pampuar, Valliyar, etc., form the major drainage system in the study area with several other creeks and brooks. Most of the beaches are erosional in nature and are enriched with workable placer mineral deposits. These are devoid of dune and in habitually expose a steep gradient in the beach face. Coconut plantation encircles most of the beaches beyond the dunes. Rich growth of mangroves and salt marshes are seen in the coast near the estuary especially in Manakudi.

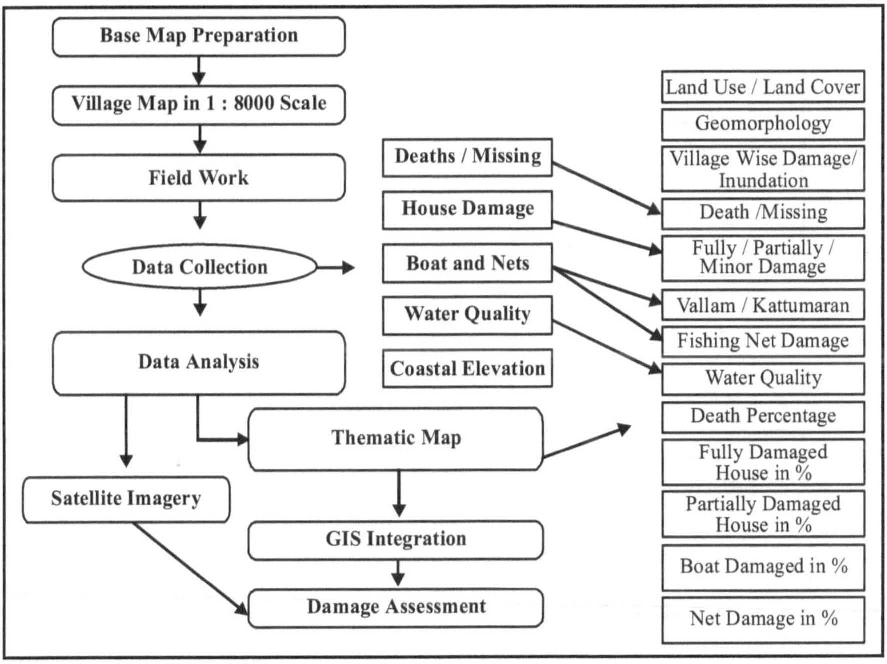

Figure 2: Flow Chart of the Methodology

MATERIALS AND METHODS

The flow chart of the methodology adopted for the present study is shown in Fig.2. The study has been initiated with the base map preparation along with the preparation of cadastral map of each coastal village on 1:8000 scale on which the satellite images have been interpolated. Intensive field work was carried out to decipher the impacts of the tsunami over natural, physical, human and biological resources. Topographic elevation and profiling were made using advanced levelling and surveying equipments following standard procedures. The field data were superimposed on the Satellite Images (LISS4 and PAN merged data) and processed in GIS environment (ArcGIS 9.1). Finally, damage assessment map was prepared.

RESULTS AND DISCUSSION

The impacts of the tsunami in the coastal region of Kanyakumari District are discussed in detail individually.

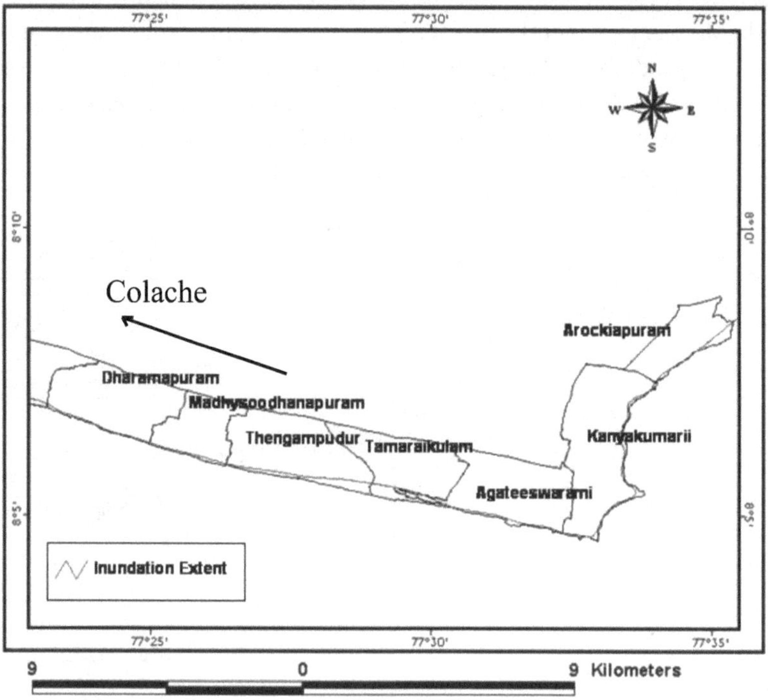

Figure 3: Map showing the Inundation between Arockiapuram and Dharmapuram

Inundation Extent (Fig.1 & 3)

Fig.3 and Table-1 show the deluge of the seawater and the run-up level inside the coast and hinterland triggered by the tsunami surge. The inundation distance and run-up level have been deciphered from the signatures of the seawater on the coastal settings and from the local enquiry. The inundation distance varies from 40 m (in Arokiapuram area) to 450 m (in Colachel area). Though Colachel ranks top in the inundation, more or less similar scenario are seen along the coastal regions of Manakudi, Pillaithoppu, Kottilpadu, etc., with the inundation ranging between 300 m and 400 m. Seawater inundation has not proceeded to a greater distance in the beaches of Chinnamuttom, Kanyakumari, Covalam, Puthenthurai, Kesavan-puthenthurai, Pozhikkarai, Periakkadu Muttom, Kadiapatanam, Chinnavilai, Periavilai, etc. Run-up level had gone to the maximum of 6 m in Colachel and almost similar situation happened along the coasts of Manakudi, Kottilpadu, Azhikkal, Pillaithoppu, etc. It has been inferred that inundation had occurred to the maximum in the beaches neighbouring the river or estuary mouths as witnessed from Colachel and Manakudi. These mouths would have, probably, facilitated the inflow of the uprushing tsunami surge far inside the coastline. There were many other Coasts which have endured only minimal inundation in spite of their proximities to river mouths. In addition to river mouths, Colachel and Manakudi had suffered maximum destruction as those locations have directly faced the refracted tsunami waves from the Southern tip of Sri Lanka. However, such waves would have been attracted more in Colachel by the river mouth neighbouring it (Raval 2005).

Table 1: Inundation Extent and Run-up Level in the Study Area

Location	Inundation limit from High tide line	Run-up level
Arokiapuram	40m	1 to 2m
Chinnamuttam	50m	1 to 2m
Kanyakumari	50m	1 to 2m
Covalam	50m	1 to 2m
Melamanakudi	400m	5 to 6m
Keezhamanakudi	350m	4 to 5m
Pallam	100m	3 to 4m
Puthenthurai	50m	2 to 3m
Kesavanputhenthurai	50m	2 to 3m
Pozhikkarai	50m	2 to 3m
Periakkadu	50m	2 to 3m
Rajakamangalam	150m	2 to 3m
Azhikkal	250m	4 to 5m
Pillaithoppu	300m	4 to 5m
Melathurai	200m	2 to 3m
Keezhamuttom	75m	3 to 4m
Melamuttom	50m	3 to 4m
Keezhakadiapattinam	50m	2 to 3m
Melakadiapattinam	100m	2 to 3m
Chinnavilai	100m	2 to 3m
Periavillai	150m	2 to 3m
Pudur	250m	3 to 4m
Kottilpadu	300m	4 to 5m
Colachel	450m	5 to 6m

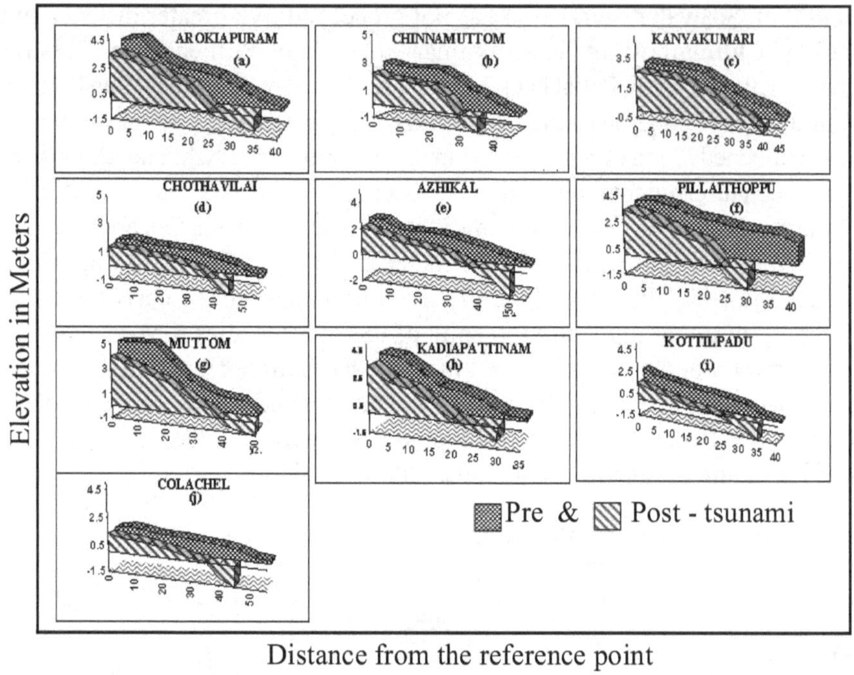

Figure 4: Impact of tsunami on the Coastal Geomorphology along the Study Area

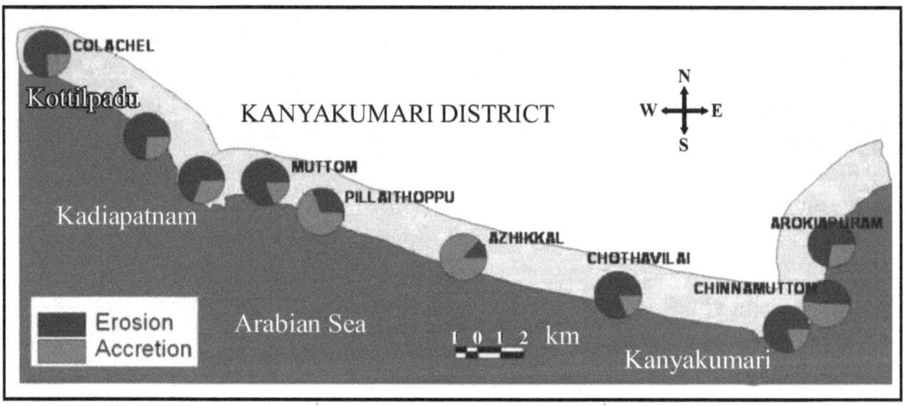

Figure 5: Erosion and Accretion of Beach Volume due to tsunami along the Study Area

Alteration in the Coastal Landforms

The authors have been already monitoring the beaches continuously in the study area for a longer period and hence could have the beach profile records for both pre and post-tsunami periods. Using such available data, the degree of erosion and accretion due to tsunami surge were calculated and the overall changes of the beach volume and morphology as well (Fig.4). The analyses reveal that the morphology of the beach has been substantially altered in the littoral zone especially in the low tide area. The backshore seems to be unaltered or altered to a bearable extent only. It is further inferred that irrespective of the coastal configuration, almost all the beaches had suffered erosion (Fig.5, Table-2). The retreating waves also eroded the beach sediments from the backshore to offshore in some parts of the coast (Koji Minoura et al 1994, Nobuo Shuto 2001).

Table 2: Amount of Accretion / Erosion of Beach Volume (m^3) due to the tsunami

Location	Accretion	Erosion	Overall Result
Arockiapuram	0.77	1.55	- 0.78
Chinnamuttom	0.16	1.52	- 1.36
Kanyakumari	1.91	0.00	- 1.91
Chothavilai	1.38	0.00	+ 1.38
Azhikkal	0.91	0.10	+ 0.81
Pillaithoppu	0.49	0.15	- 0.34
Muttom	0.07	0.56	- 0.48
Kadiapatanam	0.17	0.80	- 0.63
Kottilpadu	0.45	0.06	- 0.38
Colachel	0.24	0.30	- 0.06

Physical Damages

The details of all the physical damages caused by the tsunami surge including loss of human lives, houses and properties like boats, fishing nets, etc., are shown in Fig.7 and Table-3. From the same it is observed that casualty and settlement damage had occurred in the beaches of Manakudi, Kottilpadu, Colachel, Azhikkal where the inundation occurred towards the hinterland (Chandrasekar 2005, Narayana et al 2005). The coasts of Kanyakumari, Covalam, Pallam, Muttom, Kadiapatanam, Periavilai, Pudur, Rajakkamangalam and Azhikkal have witnessed severe casualties, house damages and fishing gears due to direct exposure to the waves and settlements located right in the CRZ (Coastal Regulation Zone). These coasts have witnessed severe damages in spite of low inundation.

Changes in Water Quality (Fig.6)

The tsunami has changed the groundwater into saline in certain areas. The level of pH and TDS (Total Dissolved Solids) in some of the coasts like Manakudi and Colachel were modified. It has been estimated that the pH level varied between 8.1 and 8.7 (Fig.6a) and the TDS varied between 2000 to 7000 mg/l (Fig.6b) along the coast. Local enquiry revealed that the quality of groundwater has become almost like seawater.

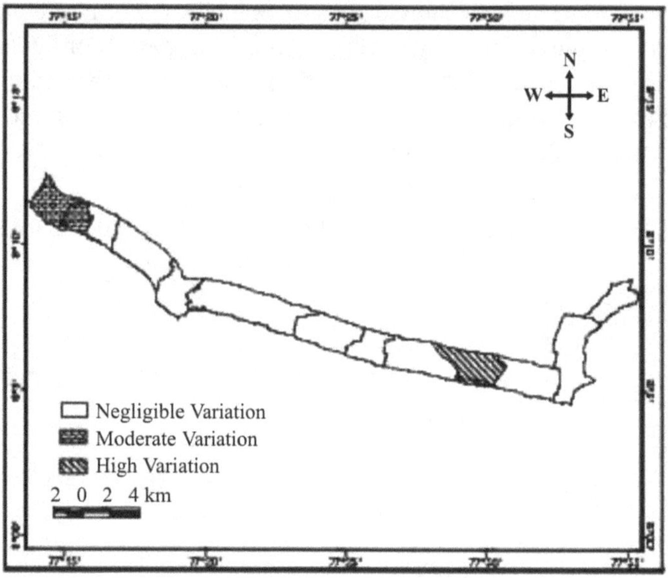

Figure 6 (a): Impact of tsunami in the pH Level along the Study Area

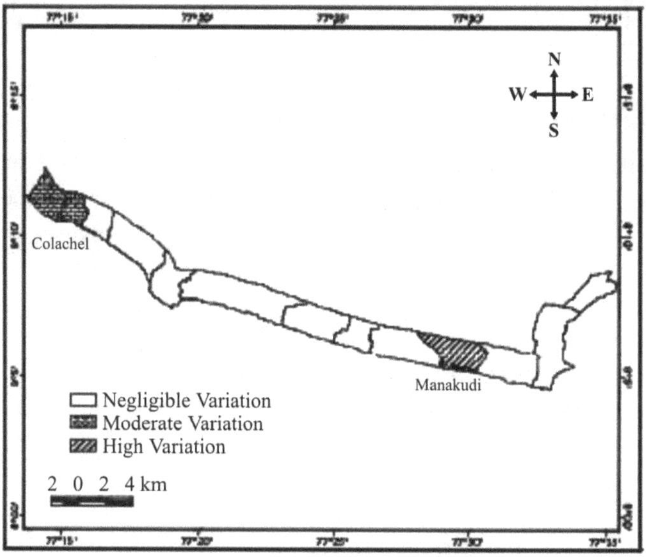

Figure 6(b): Impact of tsunami in the TDS Level along the Study Area

Biological Damages

As it is beyond the scope of any satellite imagery to delineate the minor changes in the coastal vegetation like salt marshes, mangroves, etc., due to the tsunami surge, only the observations during field work has been taken into consideration to decode the biological damages in the study area and it has been observed that excluding the negligible damage to the mangroves of Manakudi estuary (Barbara Keating et al 2004, Glenda Besana et al 2004) and to the salt marshes of some of the beaches like Colachel, Chothavilai, etc., no significant changes are seen in the study area.

Table 3: Loss of Life and Properties due to the tsunami

Location	Human Death	Houses	Boats	*Fishing Nets*
Arockiapuram	0	0	0	0
Chinnamuttom	0	0	0	0
Kanyakumari	0	0	675	900
Covalam	0	35	150	250
Keezhamanakudy	34	295	427	200
Manakudi	139	400	600	400
Pallam	6	13	540	1000
Puthenthurai	0	0	0	0
Kesavanputhenthurai	1	1	21	135
Pozhikkarai	0	0	31	25
Periakkadu	0	0	25	10
Rajakkamangalam	2	25	102	40
Azhikkal	57	300	333	40
Pillaithoppu	0	0	0	0
Melathurai	0	0	0	0
Muttom	52	187	374	750
Kadiapatanam	31	7	276	800
Chinnavilai	3	20	150	200
Periavilai	4	25	50	70
Pudur	24	60	150	200
Kotilpadu	215	402	400	400
Colachel	227	495	1200	800

Integrated Damage Assessment

The damages assessed for the natural, physical, chemical, biological and economical aspects were integrated and integrated damage assessment map was prepared (Fig.7) following standard procedure (Timothy Walsh et al 2000, Joel Bandibas et al 2003, Papathoma et al 2003, Fumihiko Imamura 2004, Chandrasekar and Loveson Immanuel 2005). From the same, the study area could be demarcated into three different zones of vulnerability. In the inundation point of view, Colachel and Manakudi were highly affected whereas in the case of physical damages which include casualties, house damages, boats, fishing nets, etc., the Coasts like Kanyakumari, Covalam, Manakudi, Pallam, Azhikkal, Muttom, Kadiapatanam,

Kottilpadu and Colachel fall in the highly affected area. The Coasts like Kesavanputhenthurai, Chinnavilai, Periavilai and Pudur fall on moderately affected areas. As far as the changes in the beach sediment volume is concerned, excluding the Coast of Chothavilai and Azhikkal, all the other Coastal segments had endured significant erosion. The tsunami surge had not shown any partiality in amending the coastal geomorphology and beach profiles; especially the intertidal zones were considerably affected. Except notable variations in the pH and TDS level of groundwater along some of the Coasts like Manakudi and Colachel, the water quality has not been affected beyond bearable limit in other areas of the study region. Fortunately, the coastal vegetations like coconut plantations, mangroves, salt marshes, etc., were least affected by the inundated seawater but for minimal damage to the mangroves and salt marshes of Manakudi, Colachel and Chothavilai. Having incorporated the above parameters, the overall impact of the tsunami waves along the study area has been prepared which shows that the Manakudi and Colachel Coasts were the highly affected areas whereas Kanyakumari, Pallam, Azhikkal and Muttom area find their place in the moderately affected area. Remaining other Coasts like Arokiapuram, Chinnamuttom, Covalam, Puthenthurai, Kesavanputhenthurai, Pozhikkarai, Periakkadu, Rajakkamangalam, Pillaithoppu, Melathurai, Chinnavilai, Periavilai and Pudur were identified as less affected areas.

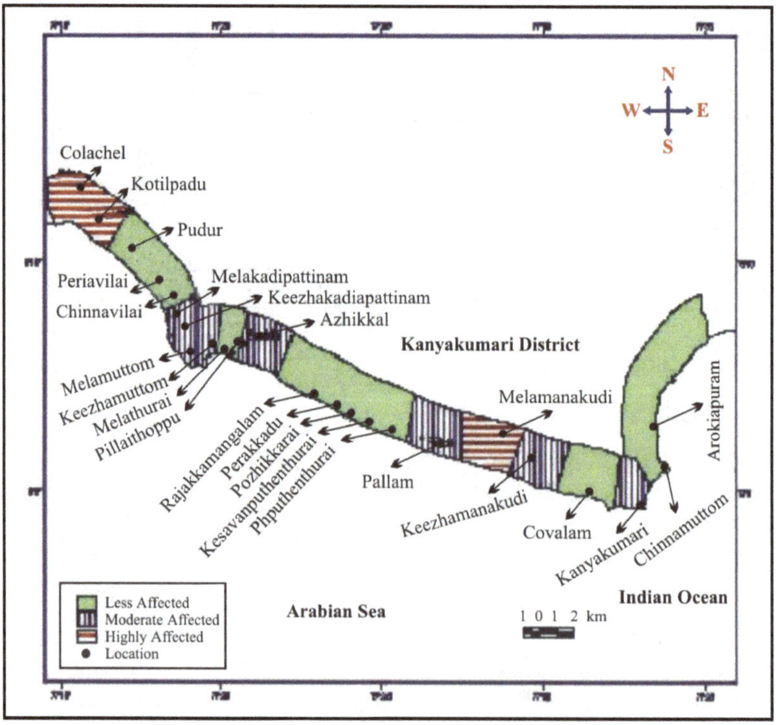

Figure 7: Integrated Damage Assessment Map of the Study Area

CONCLUSIONS

The tsunami damage assessment carried out for the study area indicates the varied impacts of the tsunami waves in different resources.

The evaluation of the coastal landforms before and after the event reveals that the uprush of the tsunami surge had brought enormous amount of sediments but the backwash had altered the beach profile especially at the intertidal zone. The estimation of beach volume discloses the fact that in spite of considerable accretion in some of the beaches the resultant sand budgeting indicates erosion.

The appraisal of the physical damages reveals that the casualty occurred more in the areas of high inundation. The buildings and the money-spinning properties like boats and fishing nets have been damaged brutally even in the beaches of minimal inundation.

It is well evident from the chemical assessment that excluding the increase of pH and TDS levels of groundwater in some of the coastal zones where there was much inundation, most of the coastal zones were found to be less affected.

The biological damage assessment proves the fact that the tsunami surge could not succeed in affecting the coastal vegetation such as salt marshes, mangroves, coconut plantations except in some of the estuaries like Manakudi with negligible degree of damage. Though the inundated seawater flooded the coconut plantations, the net damages are not any note worthy.

The damage assessment map provides a panoramic view of the destruction rendered by the tsunami surge along the study area and urges the need of proper hazard management system especially along the brutally affected Coast. As a conclusion it can be said that most of coastal zones were highly affected and the devastation caused by the tsunami surge on the coastal settings and human community may probably take some years to come back to the pre-event scenario.

ACKNOWLEDGEMENT

The authors are extremely thankful to the Department of Science and Technology (DST), New Delhi for providing financial support to undertake the study. They also thank Dr. Bhoop Singh, Director, NRDMS Division, DST for his moral support extended during the project.

REFERENCES

Barbara Keating, Franziska Whelan and Julie Bailey Brock. (2004) Tsunami Deposits at Queen's Beach, Oahu, Hawaii – Initial Results and Wave Modelling. Science of Tsunami Hazards, v. 22(1), pp. 23 – 43.

Chandrasekar, N. and Loveson Immanuel, J. (2005) GIS Supported Categorisation of Tsunami Experienced Beaches along the Southern East Coast of India: Usage in Mitigation Activities. In: Proc. Natl. Seminar on GIS Application in Rural Development, Hyderabad, India. pp. 349-362.

Chandrasekar, N. (2005) Tsunami of 26th December 2004: Observation on Inundation, Sedimentation and Geomorphology of Kanyakumari coast, South India. In: Papadopoulos and Katake (eds)., Proc. 22nd Intl. Tsunami Symposium, Chania, Greece. 27, pp. 22-28.

Fumihiko Imamura. (2004) Risk Evaluation and Real Time Information for Tsunami Disaster Mitigation. Proc. Workshop on Wave, Tide Observations and Modelings in the Asian-Pacific Region, pp. 93-100.

Glenda Besana, M., Masataka Ando and Hannah Mirabueno, M. (2004) The May 17, 1992 Event: Tsunami and Coastal Effects in Eastern Mindanao, Philippines. Science of Tsunami Hazards. v.22(2), pp.61-68.

Joel Bandibas, Koji Warita and Hirokazu Kato. (2003) Interactive Presentation of Geological Hazard Maps Using Geohazard view. Journal of Natural Disaster Science.v.25 (2), pp.75-83.

Koji Minoura, Shu Nakaya and Masao Uchida. (1994) Tsunami Deposits in a lacustrine sequence of the Sanriku coast, northeast Japan. Sedimentary Geology, v. 89, pp. 25-31.

Narayana, A.C., Tatavarti, R. and Mudrika Shakdwipe. (2005) Tsunami of 26th December 2004: Observations on Kerala Coast. Journal of Geological Society of India, v. 65(2), pp. 239-246.

Nobuo Shuto. (2001) Tsunami Induced Topographical Change Recorded in Documents in Japan. ITS 2001 Proceedings, Session, v.3(3-6), pp.513-522.

Papathoma, M., Dominey-Howes, D., Zong, Y. and Smith, S. (2003) Assessing Vulnerability, an example from Herakleio, Crete. Natural Hazards and Earth System Sciences, v. 3, pp. 377 – 389.

Raval, U. (2005) Some Factors Responsible for the Devastations in Nagapattinam Region due to Tsunami of 26th December 2004. Journal of Geological Society of India, v. 65(5), pp. 647 – 649.

Timothy Walsh, J., Charles Caruthers, G., Anne Heinitz, C., Edward Myers, P. Antonio Baptista, M., Garnet Erdakos, B. and Robert Kamphaus, A. (2000) Tsunami Hazard Map of the Southern Washington Coast: Modelled Tsunami Inundation from a Cascadia Subduction Zone Earthquake. Geologic Map GM-49. Report. Washington Division of Geology and Earth Resources. pp.12.

TSUNAMI –
ECOSYSTEM

GEOMATICS IN TSUNAMI
S.M. Ramasamy, C.J. Kumanan, Brig. R. Sivakumar and Bhoop Singh (eds.), pp. 147-153
New India Publishing Agency, New Delhi (India)
E-mail : spjain_niph@rediffmail.com Web: www.bookfactoryindia.com

11

PHYSICO-CHEMICAL PERTURBATION IN COASTAL SYSTEMS BY TSUNAMI EVENTS

A. Rajendran and R. Arthur James

Abstract

The seaquake on 26th December 2004 with its epicenter at Sumatra triggered a tsunami, which had a major impact on South East Asian countries. During that incident, the average height of the tsunami wave was 7-10 m and the penetration of seawater was up to 1-1.5 km. The surface water resources meant for irrigation and other drinking purposes got affected by the ingress of seawater in many areas. The direct environmental impacts of the tsunami have more impact along the entire east coast and caused more damages in coastal systems. The physical and chemical perturbations caused in the coastal zone by the tidal waves are discussed. In order to assess the environmental impact on coastal system, it is suggested that significant physical and biogeochemical factors could be framed to quantify the problems.

Keywords: Tsunami, Physical and Chemical Alterations, Coastal zone.

INTRODUCTION

Tsunami or tidal wave is a destructive wave that sweeps in from the sea as a huge tide. However tidal waves are not related to true tides. Instead they are caused by undersea earth movements called seaquakes or by hurricanes, cyclones or other huge storms at sea. Tidal waves caused by seaquakes are called seismic sea waves. Earthquake tremors can travel 970 km per minute, while a tsunami travels 500 to 800 km per hour in open waters. The exact speed depends on the depth of the water. Most tsunamis go undetected under the sea. However, a tsunami may form a wall of water more than 30 meters high when it approaches shallow water near shore. Tidal waves caused by large storms at sea are called storm surges. The size of a storm surge depends on the wind velocity, the duration of the storm, the sea distance over which the wind blows and the barometric pressure (Vetter 1974). Storm surges are especially destructive if they hit the shoreline during high tide period. In 1970, a cyclone and storm surge struck Bangladesh killing about 266000 people. Another cyclone and storm surge struck Bangladesh in May 1985 killing about 10000 people. Waves are distinguished from one another by different properties (Table-1)

Table 1: Various properties of waves in the Ocean.

Sl.No	Wave	Period	Wavelength	Wave type	Cause
1.	Capillary wave	< 0.1 sec	< 2cm	Deep to Shallow	Local Winds
2.	Chop	1- 10 sec	1-10m	Deep to Shallow	Local Winds
3.	Swell	10 – 30 sec	up to hundreds of m	Deep or Shallow	Distant storm
4.	Seiche	10 min – 10 hr	up to hundreds of km	Shallow or intermediate	Wind, Tsunami, Tidal resonance
5.	Tsunami	10 -60 min	up to hundreds of km	Shallow or intermediate	Submarine disturbance
6.	Tide	12.4 – 24.8 hr	up to thousands of km	Shallow	Gravitational attraction of sun and moon
7.	Internal wave	Min to hr	up to hundreds of m	Deep to Shallow	Disturbance at pycnocline

In earlier times, seismic ocean waves were called tidal waves, incorrectly implying that they were related the tides. To eliminate this confusion, scientists worldwide decided to use the Japanese word "*tsunami*" to refer to seismic ocean waves. A sudden shifting of the sea bed caused by explosive volcanic eruptions, earthquakes or submarine slumping can cause the water over the area affected to drop and generate a tsunami – a series of low waves with long periods and long wavelengths. These surface waves travel radially away from their point of origin at remarkable speeds. Because of wavelength greater than 100 km, waves in a tsunami are shallow water waves over most parts of the ocean, except over deep-sea trenches, where they are intermediate-water waves. In the open ocean, tsunamis travel at a rate of around 760 km per hour, a speed comparable to that of a modern jet aircraft. Ofcourse, in shallow water, they slow down appreciably, because of the water depth that controls their speed.

Despite these incredible speeds, tsunamis pose no danger to vessels in the open ocean, because energy and not mass is traveling at this speed. Infact sailors on the deep sea would not even notice the passage of a tsunami beneath their vessel, the regular ocean swell would hide the presence of these flat, low seismic waves, which in the open ocean are rarely higher than a meter or two. However, the waves in a tsunami can grow to a height of much greater than 10 meters, when they reach the shallows of a shoreline. Then they flood the coast, sometimes with catastrophic results, including widespread property damage and loss of life.

The earthquake on 26th December 2004 with its epicenter at Sumatra triggered a tsunami, which had a major impact on South East Asian countries. During that incident, the average height of the tsunami wave was 7-10 m and the penetration of seawater was up to 1-1.5 km. Many of the natural ecological resources like coral reefs, mangroves, forests, coastal wetlands, vegetation, sand dunes and rock formations, animal and plant biodiversity and groundwater are altered along the coastline. This particular incident made a warning alarm and hence there is a need to understand this natural phenomena's behavioral strength and their orogeny. Typically, before a tsunami, the sea will recede from the coast, exposing part of the seabed. If the slope is shallow, this recession can exceed 800 m. There are three stages of destruction during this process:- i) inundation ii) wave impact on structures and iii) erosion. Strong, tsunami-induced currents lead to the erosion of foundations and the collapse of bridges and seawalls. Additionally flotation and drag forces move houses and overturn railroads etc. In the past history of Asian records, there are three major incidents occurred in the Indian subcontinent (Table-2).

Table 2: Past Tsunami records in India.

Date	Event	Impact
August 1883	Explosion of the Krakatoa Volcano in Indonesia	East coast of India was affected, 2m tsunami recorded at Chennai
June 26, 1941	Earthquake of 8.1 Richter scale at Andaman archipelago	East coast of India affected, but no clear record on tsunami waves
November 27, 1945	Massive earthquake with 8.5 Richter scale, 100km south of Karachi	West coast of India was affectedwith 12m tsunami felt at Kandla

Source: USGS website

Beyond the heavy toll on human lives, the Indian Ocean earthquake has caused an enormous environmental impact that would affect the region for many years to come.

SURFACE AND SUBSURFACE WATER

The surface water resources meant for irrigation and drinking would get affected by the ingress of seawater in any area. Coastal freshwater aquifers are the major sources of drinking water in coastal areas. The tsunami tidal waves transport large volumes of seawater into inland water bodies and also create large tidal pools of seawater which percolates into coastal freshwater aquifers and salinise them. In the case of East coast of India, the massive quantity of seawater that inundated the coastal agricultural lands for 0.5 km to 2.0 km area inland and due to reasons of poor drainage stood for few days affecting the quality of groundwater (Singh et al 2005). The surface fresh water stored in ponds got contaminated due to subsurface seepage of seawater. Direct seawater intrusion into pond/ water storage tanks will bring damage by deposits of silts with sand as tidal wave bring sand and silt along from the coast as well as sea surface.

Tsunami waves can cause numerous direct impacts though it vary considerably in terms of extent and severity between locations, slope and distance from the sea. They are:

- Salinisation of shallow wells/surface freshwater and ground water.
- Silt and grey deposits; and sand casting on agricultural land.
- Salt deposition /salinisation of cultivable lands.
- Changes in land use brought about by loss of agricultural lands due to permanent water logging.
- Sedimentation of drainage channels, creeks' mouths, lagoons and waterways.
- Effect on fishery resources and coastal aquaculture.
- The initiation of trash and debris.
- Effect on coconuts, other plantation crops and coastal shelter belts.
- Sea erosion or sea line ingress to habitat.

SEA SEDIMENT DEPOSITION

In many areas, sediments of fine grey layer to greyish brown layer deposits of varying depths (5 cm to 35 cm) was observed in the low lying areas. Residual high content of salt is in the layers of clay and silt left behind by the tsunami waves. These layers can be easily identified by cracks that spread across the surface of the soil. The top silt covering the salt affected soil further reduces the permeability of the soil. In order to improve such situation removal of salt by surface scraping or flush or leach by huge water application are alternatives to quick reclamation. Localized tilling, controlled irrigation and water application will help to keep the salt away from crop root zone. This is possible for small areas only.

ENVIRONMENTAL IMPACTS

The direct environmental impacts of the tsunami have some impact along the Sumatra-Nicobar fault line on Indian side of Andaman and Nicobar Islands. In the union territory of Andaman & Nicobar Islands many coastal wetlands have been affected specially, by the large inflow of salt water and silt load (Ramachandran et al 2005). Unconfirmed report also indicates that the post-tsunami Andaman and Nicobar Island's geo-position is modified. There has been some terrestrial and aquatic pollution, both in terms of physical debris as well as sewage and effluents from industries and chemical establishments.

The primary effect of total salinity is a reduction of water availability to roots through osmotic effects. Yields of most crops are substantially reduced due to higher TDS. Hence post- tsunami soil TDS levels have increased by 5 to 15 times compared to normal season, while soil pH levels increased marginally. In few areas there was unique impact of stagnation of sea sediments, debris and seawater. The thick slushy black deposits on the soil surface causing heavy damage to the soil structure. The post-tsunami reports revealed considerable ecological changes particularly in the severely affected areas. The data showed changes in inter tidal and sub tidal faunal assemblages. The density and diversity of phyto and zoo plankton had been reduced, while the benthic macro fauna was almost absent in all areas in the inter tidal zone. There were vast changes in seafloor topography and shoreline, mostly transgression (50-250 m) and regression (5-20 m) in some areas. The deposition of heavy minerals such as garnet and illmanite was also reported along the coastal tract up to 3-20 m away from the shoreline. No significant impact was noticed on the coral and mangrove ecosystems, but due to strong waves few table corals (Acropora cytheria) were tilted, and branching corals (Acropora intermedia, Acropora nobilis) were broken . Some mangrove plants were uprooted in Andaman and Nicobar Islands and the only major impact was sedimentation, which blocked free movement of water (Ramachandran et al 2005). In the case of other coastal plantations (casuarina, coconut and palm), the leaves became dry, otherwise, they withstood the tsunami impact except uproot of few coconut trees in some places. Coral reefs, mangroves, casuarina, coconut and palm acted as natural barriers and minimized the impact.

EFFECT ON THE CHEMISTRY OF COASTAL WATERS

The immediate and phenomenal effect by the tsunami event on coastal waters would be the change in salinity, temperature, pH and oxygen contents. A drop in temperature followed by an increase in salinity drastically affects the coastal living systems, from bacteria, coral reefs to fishes. The oxygen concentration suddenly decreases and a shift in pH and conductivity would at least partially destroy the ecosystem.

Intricate chemical and geochemical changes are also possible. Dissolved organic matter with high molecular weight and slow oxidation rate would be added to the surface waters. These resilient organic compounds would slowly strip away the trace elements and micronutrients from their soluble condition by coagulation and/or chelation. There might be an alternate geochemical process happening due to the change in quality and quantity of organic constituents of seawater.

Still more drastic change could be expected in the biogeochemical process in the sediment – water interface in the continental shelf. Change in the rate of transport of elements between sediment and pore water is imminent due to the change in carbonate/ bicarbonate chemistry of neritic waters.

In Sri Lanka's coastal areas, for example, artisanal fishery is an important source of fish for local markets and industrial fishery is the major economic activity, providing direct employment to about 250 000 people. In recent years the fishery industry has emerged as a dynamic export-oriented sector, generating substantial foreign exchange earnings. Preliminary estimates indicate that 66 percent of the fishing fleet and industrial infrastructure in coastal regions have been destroyed by the wave surges, which will have adverse economic effects both at local and national levels. But some economists believe that damage to the affected countries' economies will be minor because losses in the tourism and fishing industries are a relatively small percentage of the GDP. However, others caution that damage to infrastructure is an overriding factor. In some areas, drinking water supplies and farm fields may have been contaminated for years by salt water from the ocean.

Both the earthquake and the tsunami may have affected shipping in the Malacca Straits by changing the depth of the seabed and by disturbing navigational buoys and old shipwrecks.

Salted-over soil becomes sterile, and it is difficult and costly to restore for agriculture. It also causes the death of plants and important soil micro-organisms. Thousands of rice, mango and banana plantations in Sri Lanka were destroyed almost entirely and will take years to recover. The United Nations Environment Programme (UNEP) is working with governments of the region in order to determine the severity of the ecological impact and how to address them.

In order to assess the environmental impact on coastal system, the following factors could be useful to quantify the problem:-

- Salinity levels in the topsoil, subsoil, sediment and groundwater, the debris deposited
- The type of sediment, its thickness, texture and consistency, drainage properties and chemical composition,

- Interactions between sediment and topsoil depending on the type of topsoil,
- The availability of irrigation and drainage channels and the damage to surrounding irrigation infrastructure, and
- In coastal areas the degree to which tidal flows can be controlled.

The appropriate solution could be possible if suitable eco-friendly guidelines such as from Coastal Regulation Zone (CRZ) Authority can be implemented to save the ecosystem. Also afforestation of deforested areas, including mangroves and resettlement of coastal population in safer zones should be encouraged.

REFERENCES

Ramachandran, S., Anitha, S., Balamurugan, V., Dharanirajan, K., Irene Preeti Divien M., Senthil Vel, A., Hussain, I.S. and Udayaraj A. (2005).. Ecological impact on tsunami on Nicobar Islands (Camorta, Katchal, Nancowry and Trinkat). Current Science, v. 89 (1), pp. 195-200.

Richard C. Vetter (1974). "Oceanography" Voice of America, Forum Series, 430p.

Singh, V.S., Nandakumar, M.V., Sarma, M.R.K. and Dimri, V.P (2005). Changes in groundwater regime at Neill Island (South Andaman) due to earthquake and tsunami of 26 December 2004. Current Science, v.88(12), pp.1984-1987.

GEOMATICS IN TSUNAMI
S.M. Ramasamy, C.J. Kumanan, Brig. R. Sivakumar and Bhoop Singh (eds.), pp. 155-166
New India Publishing Agency, New Delhi (India)
E-mail : spjain_niph@rediffmail.com Web: www.bookfactoryindia.com

12

26, DECEMBER 2004 GREAT TSUNAMI: IT'S CHARACTERSTICS AND COASTAL RESPONSES IN CAR NICOBAR ISLAND, INDIA

Arun Kumar, R. K. Chingkhei and Bhoop Singh

Abstract

The Car Nicobar Island is one of the several islands in Indian Ocean which was severally damaged by the Great tsunami on 26 December 2004. The great tsunami event caused massive devastation, in the form of loss of life and property in south and south East Asia. The objectives of the study are to assess the changes in the coastal features of the Car Nicobar Island using various indicators such as the tsunami height, run- up elevation, water flow direction, erosion, sand deposition and coastal subsidence. Andaman & Nicobar Islands are the summit of the submarine mountain range lying along the great tectonic suture zone extending from Eastern Himalaya to Sumatra and Lesser Sunda via Myanmar border and Arakan hills. The IRS P-6 satellite data has been found to be very useful in the present tsunami assessment studies. The digital analysis of the satellite data was done using the image processing software (ERDAS IMAGINE 8.4) and DTM to quantify the tsunami height (15 m), run up elevation (2-19 m), inundation (295-1203 m) and coastal responses. The settlements and coconut plantations have been severely damaged in the island, which were assessed using GIS analysis of the pre and post-tsunami satellite data.

Keywords: Tsunami, Wave Height, Run-up Elevation, Inundation, Coastal Responses, Remote Sensing, GIS.

INTRODUCTION

The Car Nicobar Island is one of the several islands which were severally damaged by the Great tsunami of 26 December 2004. The origin of tsunami was due to a series of undersea earthquakes, the largest being 9.3 M. in Richter scale. The direct consequence of this Great Earthquake, that ruptured the sea floor to a width of up to 100 km in places, was the displacement of a huge volume of water that translated into tsunami of colossal proportion. The great tsunami event caused the devastation and loss of life in South and South East Asia including the Andaman and Nicobar Islands. The geological formations represent the period of sedimentation from Cretaceous to Sub Recent period. The Nicobar Island is a part of a great island arc created by the collision of the Indo- Australian Plate with Eurasia. The collision lifted the Himalaya and most of the Indonesian islands, and created a long arc of highlands and islands, which includes the Arakan Yoma range of Myanmar, the Andaman - Nicobar island, and the islands off the west coast of Sumatra, including the Banyak Islands and Mentawai Islands (http://www.Forest.nic.in/frstisland1. htm). These islands are the summits of a submarine mountain range lying on the great tectonic suture zone extending from the Eastern Himalayas along the Myanmar border to the Arakan and finally Sumatra and Lesser Sundas. The ages of geological formations range from Cretaceous to Sub - Recent period. The surface deposits of gravel beds and raised soil covers that are of recent origin seem to be of Holocene age (<10,000 yrs). The present configuration of these islands took shape only about 26 million years ago.

These islands have a tropical climate and the temperature ranges from 22° C to 28°C. Rainfall is heavy due to annual monsoons and measures around 3000 to 3800 mm. The physiography of these islands is characterised by undulating topography and intervening valleys. There is no major perennial fresh water river in these islands, but for a few rain fed streams. The coast lines of these islands are wavy with large number of bays and lagoons. Soil cover is thin, varying from 2m to 5m. The coastal flats have an admixture of sand, silt and clay with fine fragments of coral lime. The vegetation of the Nicobars is typically divided into the coastal mangrove forests and the interior evergreen and deciduous Tropical and subtropical moist broadleaf forests.

The present study was conducted in Car Nicobar island to assess the damages of tsunami and to find out the geological evidences for tsunami characterization using Remote Sensing and GIS. The objectives of the study are to assess the changes in various coastal features of the island using different indicators such as the tsunami height, run-up elevation, water flow direction, erosion, sand deposition and coastal subsidence.

GREAT EVENTS OF 26 DECEMBER 2004

The Great Earthquake of December 26 2004 that occurred off the west coast of Northern Sumatra took place at the interface between India and Burma plates, the Burma plate has been referred as Andaman/Nicobar Ridge that acts as a small tectonic plate (Curray et al 1982). The event was considered as a mega thrust earthquake referring to the large cracking of the plate boundary. Such mega thrust earthquakes often generate large tsunamis that can cause damages over a much wider area than is directly affected by ground shaking near the earthquake's rupture. Tsunami was generated in the fast slip area (first 650 km at taut length) and the waves propagated in all directions. As a result, strongest waves hit the coasts of Indonesia, Thailand and other nearby areas (Andaman & Nicobar Islands) which are closely located to the east of the epicenter. The devastation in the Nicobar Island was due to a 10-15 m high tsunami following the 2004 Indian Ocean earthquake. At least 3000 people were believed to have been killed on the disaster.

Historical data does not provide a very realistic figure on the earlier tsunami events in the island. However, 31, Dec1881 tsunami triggered by an earthquake of 7.9M has been recorded by the National Geophysical Data center, NOAA, USA (www.ngdc.noaa.gov/nmdc/servlet/showdatadatasets). But the details about the tsunami hazards of the other earthquakes and volcanic eruptions of 1847 (Earthquake), 1883 (Volcanic eruption) and 1941 (7.7 M) are not know. The possible clues can be explored by trenching the beaches, study their profiles and date them by TL/OSL technique to ascertain the historical and pre-historical tsunamis in the region.

MATERIALS AND METHODS

Materials

Software - ArcView, ERDAS IMAGINE etc. Data - Survey of India (SOI) topographic Maps, Satellite images- IRS-IC LISS III (24 Feb. 1999), IRS-P6 (16 Feb. 2005 & 01 Feb. 2005), GPS- Garmin etrex & Garmin vista, Sony Handycam, Sony Digital still camera, High precision Oregon scientific Altimeter, Measuring tapes, Staff etc.were used in the study.

Methodology

The IRS P-6 satellite payload has been found to be very sensitive system in this regard. The tilting of camera to see the affected areas from an easterly path of the satellite has been proven to be of high practical utility combined with wide view of these sensors. The digital analysis of the P6 satellite data was done by using the image processing software (ERDAS IMAGINE 8.4). Various steps have been followed in order to arrive at best results in precise delineation of the different tsunami affected areas. It is observed that the image enhancement techniques with brightness contrast and break points have potentials in delineating the inundation of tsunami water in the islands. The natural colour simulation was done to map coastal features precisely.

The coastal mapping was done through on screen digitization. The estimation of wave height, run-up elevation, and delineation of coastal erosion and impact of damages were done using the DEM/DTM. In the present study the DTM was produced from topographic maps and SRTM data with the cell size of 22 meters. The field data of the various locations that were collected with the help of handheld GPS (Garmin etrex & vista) were used as overlay on the Digital terrain model to verify the tsunami wave height as well as the distance from the sea.

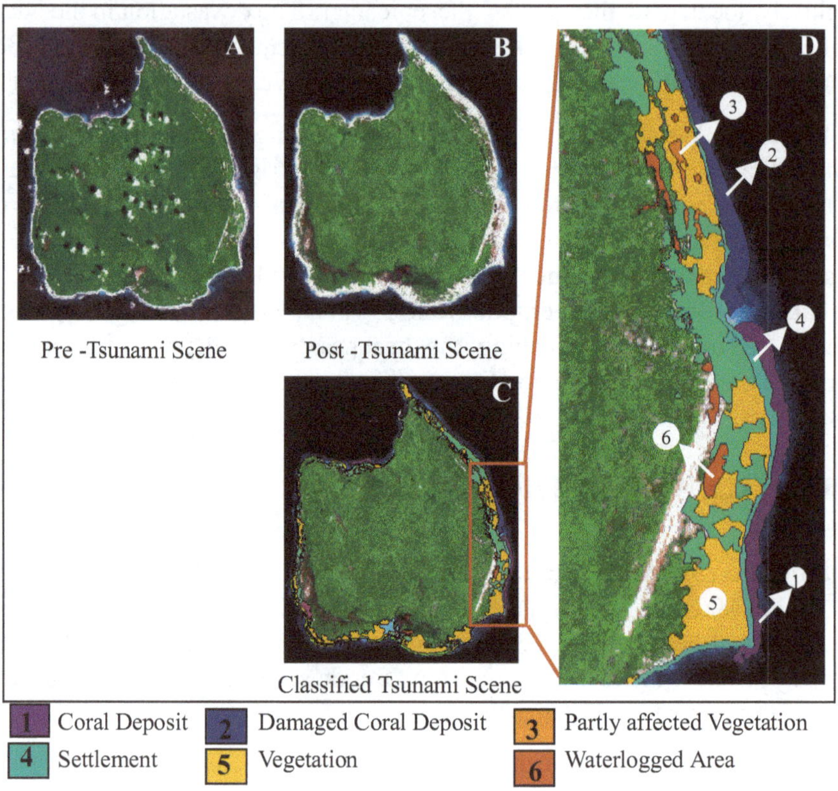

Figure 1: Stages of Mapping a Tsunami.

RESULTS & DISCUSSION

There are number of observations by various workers on the characteristics of past tsunami hazards (Yalciner et al 1999, 2000 and 2001, Altinok et .al. 1999 and Musaoglu 2000). In the present study tsunami hazards were delineated and assessed by the various observations during investigations and the following results were arrived in:

Tsunami Wave Heights

The tsunami wave heights were measured based on the satellite images and the DEM generated using SOI contours and 1m SRTM data with a vertical resolution of +/-1m. In the present study, DTM was produced from topographic maps and SRTM data which has a cell size of 22 meters. The field data of the various locations that were collected with the help of handheld GPS (Garmin etrex & vista) were used as overlay on the Digital terrain model to verify the tsunami wave height as well as the distance from the sea. In order to assess the tsunami height on all the sides of the island, four strategic locations were selected viz. North eastern (NE), North western (NW), South eastern (SE) and South western (SW) part of the island. The maximum tsunami wave height was 15m on the south east and minimum was 0.7 m with a distance from the shoreline of 400 m and 368.58 m respectively. The photograph taken from a low altitude aircraft also depicts the evidences of tsunami wave heights and the distance from the shore on the south eastern part of the island.

The various indicators such as wave heights, water stains on the buildings, salt burnt trees and broken branches of the trees and debris on the trees (Fig.2) were also used to measure the run-up elevation in the field survey. The manual methods were used to assess the heights. The southeast facing coastlines near Malacca area were struck by the highest waves of more than 15 m. Waves that hit the north facing coastline of Mus and Passa were low, (about 8-10 m high), but the low-lying nature of the land allowed those waves to penetrate far inland. It was observed during the survey that many two storied buildings were inundated by tsunami waves and trees had the debris brought by tsunami (Fig.2).

Figure 2: Debris on trees indicate tsunami height

The hypothesis of tsunami height was made by measuring wave heights at intermediate points along the coastline and to collect additional data on sediment-deposit profiles (USGS, http://walrus.wr.usgs.gov/tsunami/sumatra05/ heights.html). Efforts were focused mainly around the very south-east end of island around Malacca, Kakana, Chukchucha, Lapathy and Kimus to collect the data. Wave heights of 15 m (50 ft) at those sites suggest that the tsunami waves might have been 10 to 15 m high along the entire stretch of coast from Malacca to Kimus.

Run Up Elevation

In general, the extent of vertical run-up of seawater during tsunamis depends on earthquake parameters, geographical locations, velocity of tsunami waves and their frequency, near shore bathymetry, beach profile, land topography etc. Due to these parametric variations in various coastal areas including the islands, the run-up levels and landward penetration characteristics of seawater are location specific and vary significantly within a short distance and even in an island. In the case of Nicobar Island the run up levels varied from 2 m to 19m and the distance of penetration from the coast ranged from 295.87 to 1202.57m (Fig.2).

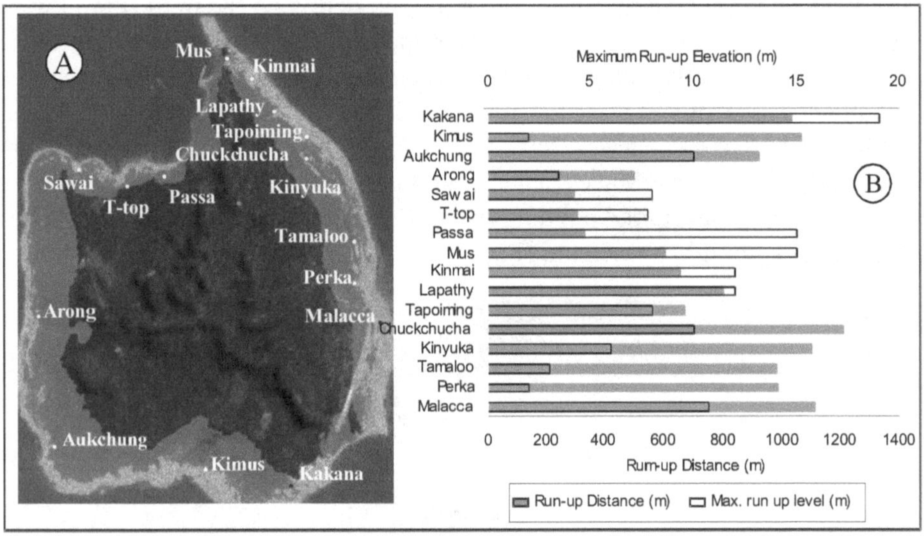

Figure 3a : Tsunami run-up model generated using the DEM/ DTM and SRTM data with a run-up elevation of 19m. Car Nicobar being low lying area which is almost flat at some places, the DEM has been exaggerated 9 times the actual to highlight the hidden terrain and to have a better perception of the model.

Figure 3b: Bar graph showing the run-up levels and their respective run-up distances.

Bilham et al (2005) have drawn preliminary conclusions on the slip pattern of 26 Dec 2004, earthquake that due to high rate of slip in the southern 650 km of the 1300 km long North –South rupture zone of 2004 Andaman-Sumatra earthquake, the principal tsunami was generated in the Sumatra area. The time lag between the earthquake and land subsidence in Car Nicobar on 26 Dec 2004 which is estimated to be 15-20 min has been interpreted as the high rate of slip in the Nicobar region resulting the generation of near-field high tsunami (19 m run- up level) in this zone. The above parameters have caused the land subsidence at Nicobar Islands due to the earthquake. The run up levels recorded at Andaman groups of Islands were different due to the low slip rate, caused by the earthquake. Similar types of diversified results were observed on 26th December 2004 tsunami affected coastal areas in Thailand, Indonesia and Sri Lanka. Run up levels varying from 0.3 to 32 m were recorded in Indonesia and from 2.5 to 10 m in Sri Lanka (Yalciner et al 2005) and 2.5 m in Seychelles. It is observed that the run up level was affected by the earthquake mechanism and resulted to different run up levels at different coastal areas. In case of Nicobar Island, the run-up levels have varied from place to place as shown in the Fig.3.

Inundation

The inundation distances in the island were so large that they were most easily measured from satellite images, where sediments deposited by the waves and vegetation killed by the saltwater are clearly visible (Fig.3a). Satellite images show that such waves that struck many such villages like Malacca, Kakana, Kimus and Mus are located SE, SW and NE parts of the islands. Items broken and bent by the tsunami waves were used to determine flow directions. The team found that the large tsunami waves flowed around natural barriers, flooding lowlying areas behind them. The inundation limit could be clearly seen on Kakana beach and Kimus beach (Fig.4).

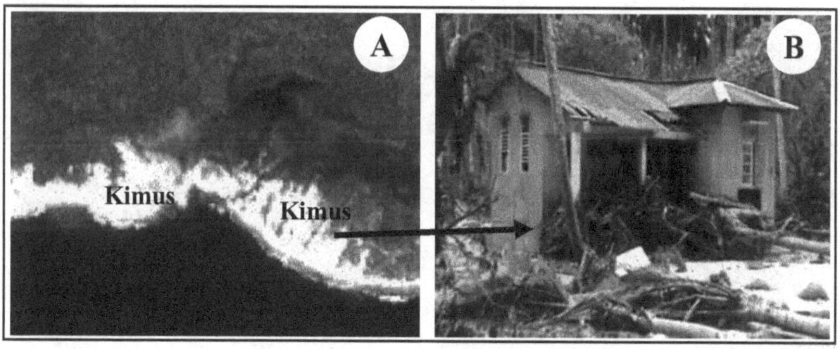

Figure 4: Inundation at Kimus beach indicated by the white colour on the satellite image (A) due to the fresh beach sand and field picture (B) indicating the debris flown by the tsunami water from beach to island and struck by a house

Tsunami Flow Direction

The Car Nicobar Island has been affected by the water inundation. The Team found that the large tsunami waves flowed around natural barriers, flooding low lying areas behind them. Flow direction has been marked by a number of indicators such as water marks on the walls of the buildings, trees, DEM, satellite imageries, items broken and bent by the tsunami waves etc. It is observed that Nicobar Island, being very nearer to the earthquake source, the tsunami water inundated from all the sides and flown towards the inland. The flow direction of the tsunami water was from all sides of the island (Fig.5).

Sediment Deposits

The team surveyed beach profiles to document erosion (common near the coast) and deposition (common inland) by the tsunami. Sand eroded from beaches probably provided much of the sand that was deposited inland. The survey team dug trenches in the tsunami deposits to measure their thickness and to examine other characteristics that can shed light on how high the waves were and how fast the water was flowing. Data from the sediment deposits will not only reveal about the recent tsunami but also give information on the ancient tsunamis, which, in turn, will help in better understanding of the tsunami history of an area. From such tsunami history the likely history and the future possible events too can be predicted. The Kakana beach and Aukchung beach are selected for the study. The sand deposit is 2.5 m at Aukchung beach and 1 m at Arong beach (Fig.6).

Figure 5: Flow of tsunami water is marked on satellite image and site pictures by arrows

Coastal Subsidence

The subsidence observed at beaches near Malacca Jetty is 0.75 m and a temple on Malacca beach is submerged by 1-1.25 m approximately (Fig.7). It is inferred from the observation that the subsidence may be the result of the earthquake (mega thrust caused the rise of sea floor and subsidence at beach). It is also predicted by the model used by USGS that the type of earthquake that caused the tsunami - a megathrust - will raise the sea floor above the fault rupture and cause subsidence near the coast. The pre and post-tsunami images have been studied for the observation of subsidence.

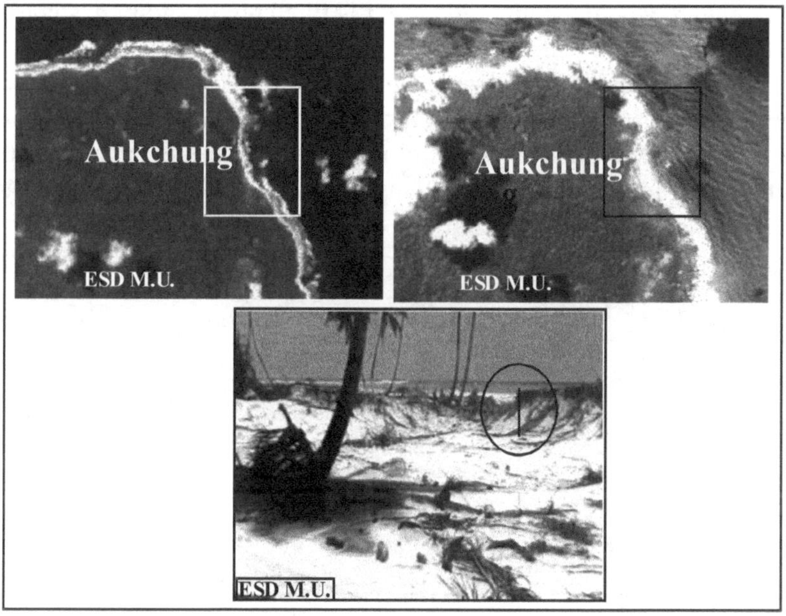

Figure 6: Showing Pre (upper left) and post (upper right) satellite images and the field photo of Aukchung beach in the 2.5 m thick sand deposits

Figure 7: A part of Malacca beach (left) and Murugan temple (right) showing coastal subsidence

Erosion and Coastal Responses

The Car Nicobar Island underwent significant modifications by the tsunami. The shoreline eroded, beach sand was carried inland and the coastal plain was flooded. Compared to the erosion and deposition of the sediments by the tsunami that occurred relatively quickly, within hours of the initial tsunami impact, coastal subsidence resulted in additional erosion and shoreline retreat during the weeks and months following the tsunami. The observations from the field studies indicate coastal erosion at Aukchung, Kimus and Malacca villages where the road is washed off, Mangroves and bridges (Fig.8) are damaged and underwater cracks extended inward where the high tide water is filled up. In addition, some parts of beaches have new sand deposits, located near Aukchung and at some places near Kakana where the beaches started rebuilding as soon as a few weeks after the tsunami, probably adding sand from nearby offshore. These reformed beaches were migrating landward through over-wash processes. Some beaches were still migrating landward, impacting roads and redevelopment plans for coastal villages. It is expected that the future shoreline retreat may continue to impact redevelopment in some areas.

Figure 8: Coastal erosion along at Aukchung Village (left) and filled up high tide water inland (right)

Assessment of the Damages

In order to assess the damages caused by the Great tsunami in the Car Nicobar Island the pre and post-tsunami satellite data were critically analysed in the GIS domain. For this, a base map was generated from the SOI topographic map of the region. The Base Map was later on updated from the field data and information obtained from the Nicobar Island Administration office. Based on this map the coastal area of the two scenes (pre and post) have been classified and vectorised using the ERDAS vector and ArcView software (Fig.1). Overlay analysis of these classified vector data was performed to find out the changes in the coastal corridor of the Island and ultimately the assessment of the damages was done from this analysis. The result of the assessment is tabulated as below (Table-1).

CONCLUSIONS

The Great Indonesian Earthquake of 26 December 2004 and the subsequent Great tsunami event has led to the wide spread devastation in the Indian Ocean coast and all islands in its vicinity. These events have created ways and means to explore the nature's mystery of natural processes / hazards in the Indian Ocean. The damages caused by these hazards are being accounted to the losses of millions of dollars and loss of property and life of people and many more homeless. The application of High resolution Remote Sensing data and GIS techniques were used to assess the tsunami hazards in the Car Nicobar Island. Maximum tsunami wave height was 15 meters in the south eastern parts of the island and minimum of 0.7 meters with a distance from the shore line of 400 meters and 368.58 metes respectively. The run up levels varied from 2 to 19 m and penetration distance from the coast, ranged from 295.87-1202.57 m on the inland. The tsunami water flowed from all the directions in to the island. A considerable part of all the existing beaches hosting mostly infrastructural, commercial and residential complexes have been fully damaged. The sand deposits at Aukchung and Arong beaches were 2.5 and 1 meter respectively. The coastal subsidence near Malacca Jetty is 0.75 m and a temple on Malacca beach is 1-1.25 meters. The car Nicobar Island underwent significant coastal modifications and over all the coastal area affected by the tsunami is 2075.26 ha. The study indicates considerable coastal erosion at Aukchung, Kimus and Malacca villages.

Table 1: Damage Assessment of tsunami impact.

Class Name	Pre-tsunami 1999 (Area in Ha)	Post-tsunami 2005 (Area in Ha)	Assessment from 1999 -2005 (Area in Ha)
Coconut Plantation	36.70	6.25	30.45
Coral Deposit	230.90	103.44	127.46
Edged Rock	126.01	8.64	117.37
Mangrove Swamp	88.91	40.00	48.91
Sand	201.52	49.63	151.89
Settlement	667.68	58.01	609.67
Sheet Rock with Coral Deposit	84.05	1.68	82.37
Shingle and Sand	127.43	89.03	38.40
Vegetation	11598.30	10776.59	821.71
Tota l Area in Ha	**13161.51**	**11133.27**	**2028.24**

ACKNOWLEDGEMENT

The financial assistance provided under the DST grant No. SR/S4/ES-135/ 3.5/2005 is thankfully acknowledged. The authors are grateful to A&N Administration for their help during the field work.

REFERENCES

Bilham, R., Engdhal, E.R., Feldl, N. and Satyabala, S.P. (2005) Partial and Complete Rupture of the Indo-Andaman plate boundary 1847-2004. Seism. Res. Lett. Draft dt. 4 March 2005 published in the internet;

http://cires.colorado.edu/~bilham/IndonesiAndaman2004_files/AndamanSRL4Mar.

http://cires.colorado.edu/~bilham/IndonesiAndaman2004_files/AndamanSRL4Mar.htm

Curray, J. Emmel, R., Moore, D.G. and Raitt, R.W. (1982) Structure, tectonics and geological history of the NE Indian Ocean. In: The Ocean Basins and Margins, The Indian Ocean, A. E. M. Nairn and F. G. Sehli, Plenum (eds), New York, v. 6, pp. 399- 450.

Musaoglu, N., Kaya, S. and Seker, D. Z. (2000) Investigating earthquake effects using GIS and remote sensing data, Proceedings of the 20th EARSEL Symposium, Version 1, pp. 235-240.

Yalciner, A.C., Synolakis, C.E., Borrero, J., Altinok, Y., Watts, P., Imamura, F., Kuran, U., Ersoy, S., Kanoglu, U. and Tinti, S. (1999) Tsunami generation in Ýzmit Bay by the Ýzmit Earthquake, Proceedings of ITU-IAHS International Conference on the Kocaeli Eartquake 17 August 1999, Ýstanbul, pp. 217-221.

Yalciner, A.C., Altýnok. Y. and Synolakis, C.E. (2000) Tsunami Waves in Izmit Bay after the Kocaeli Earthquake, "Kocaeli, Turkey Earthquake of August 17, 1999 Reconnaissance Report, published by EERI, (Earthquake Engineering Research Institute, USA), (special issue supplement to Volume 16) pp:55-62

Yalciner, A. C., Synolakis, A.C., Alpar, B., Borrero, J., Altinok, Y., Imamura, F., Tinti, S., Ersoy, S., Kuran, U., Pamukcu, S. and Kanoglu, U. (2001) Field Surveys and Modeling 1999 Izmit Tsunami. International Tsunami Symposium ITS, August 7-9, Seattle, pp. 557-563

Yalciner, A.C., Altinok, Y. and Synolakis, C. E. (2001) Tsunami waves in Izmit Bay after Kocaeli earthquake. Journal Earthquake Spectra.

Yalciner, A.C., Perincex, D., S.Ersoy, S. Presateya, R. Hidayat and McAdoo, B. (2005) Report on December 26, 2004, Indian Ocean Tsunami, Field survey on Jan 21 – 31 at North of Sumatra by ITST of UNESCO IOC, 2005.

GEOMATICS IN TSUNAMI
S.M. Ramasamy, C.J. Kumanan, Brig. R. Sivakumar and Bhoop Singh (eds.), pp. 167-177
New India Publishing Agency, New Delhi (India)
E-mail : spjain_niph@rediffmail.com Web: www.bookfactoryindia.com

13

REFLECTION OF BOXING DAY TSUNAMI ON CORAL REEF ECOSYSTEM OF INDIA

L. Kannan, T. Thangaradjou and R. Sridhar

Abstract

Coral reefs, the marine representatives of tropical forests, coupled with the sea grasses contribute considerably to the coastal productivity. Damages to these ecosystems will immediately affect the fishery productivity of the region. Due to this, the long-term ecological impact of the recent tsunami will be a serious set back to the local, national and regional economy than the immediate losses. Andaman and Nicobar reefs represent the second largest coral reef formation next to the Great Barrier Reef and are also comparable to the coral reefs of Indonesia and Philippines in terms of diversity of species and associated floral and faunal components. The recent mega – subduction of the Indian plate below the Burmese micro – plate in Andaman – Sumatra region has not only deepened the southern and uplifted the northern group of Andaman – Nicobar islands but also created a disastrous tsunami on Dec 26, 2004. While such tectonic changes are expected to cause longterm changes in the coral reefs of Andaman, Nicobar group of islands, the tsunami has damaged the coral reefs instantaneously. However, the coral reefs of the Gulf of Mannar and Palk Bay are not much affected by the tsunami by virtue of their geographic position in the shadow region of Sri Lanka.

Keywords: Indian coral reefs, Tsunami effect, Coral recovery

INTRODUCTION

Coral reefs form the world's most dynamic ecosystems, providing shelter and nourishment to thousands of marine microbial, floral and faunal components, besides protecting the coastline. This unique ecosystem is more productive because of its symbiotic association with zooxanthellae. Though they are the builders of most massive structures ever created by living beings in the world, they are very fragile and vulnerable to natural disturbances, and human activities (Venkataraman 2003). The marine representative of the tropical rain forests viz. coral reefs of the world occur in 6,00,000 km^2 and the reefs of the ASEAN region, especially from Indonesia and Philippines alone, contribute more than 30% to the world total. The coral reefs of Southeast Asia are considered to be the center for evolution of hermatypic corals (Ming 1997) and also the global center of biodiversity for reef fishes, molluscs, crustaceans, mangroves and seagrasses, but they are under high potential threat (Wilkinson 2002). Further, the Regional Marine Programme has ascertained that the coral reefs of the Indian Ocean have just begun to recover from the damage caused by the El-Nino (1997 & 1998) and hence the recent impairment caused by the tsunami can be very serious and the coral recovery may take a decade to a century (Kannan et al 2005a) in the affected areas.

Recent tsunami on the 'Boxing Day' (26 December 2004), triggered by the earthquake epicentred on the coastal region of the Sumatra island, Indonesia caused huge and immediate damages to human lives, properties and economy of the affected countries, as most of the affected nations are promoters of coastal tourism and fisheries. But many people do not really know the extent of ecological damages caused by these killer waves. These ecological damages include loss of marine living resources, especially the ecosystems like coral reefs, sea grasses and mangroves; damages to coastal littoral forests, loss of turtle and bird nesting grounds and increase in ground water salinity. All these may result in a series of interlinked impacts on the entire marine ecosystem of the Indian Ocean in the coming years. This will cause considerable imbalance in the economy of the countries in general and socio - economy of the coastal people in particular.

CORAL REEFS OF INDIA

Coral reefs in India occur along the east and west coasts, at selected places (Table-1). In India, all the major reef types are represented: fringing reefs are found in the Gulf of Mannar and Palk Bay; platform reefs are seen along the Gulf of Kachchh; patchy reefs are present near Ratnagiri and Malvan coasts; fringing and barrier reefs are found in Andaman and Nicobar islands and atoll reefs are found in Lakshadweep islands. Over 300 species of corals have been identified so far from Indian waters. Among them, the coral reefs of Andaman and Nicobar, Gulf of Mannar and Palk Bay regions have been affected by the recent tsunami directly and / or indirectly. Unfortunately, these coral reefs are harboring maximum coral diversity in the country. The Gulf of Mannar and Palk Bay embrace 128 species and Andaman and Nicobar islands have 203 species.

Though coral mining and destructive fishing have been totally banned in the coral reef areas, there are severe damages to the corals due to other man made causes which are largely linked with the socio-economy and life style of the local people. For example, collection of molluscs in the Andaman and Nicobar and Gulf of Mannar regions, seaweed harvesting and pearl oyster collection in the Gulf of Mannar and Palk Bay reef areas exert a lot of pressure on the reefs of these regions. While coral reefs of the Gulf of Kuchchh are largely affected by the coastal pollution caused by the industries, corals of some remote islands of the Andaman and Nicobar and Laskshadweep islands are facing threats due to the coral and beach sand mining for construction purposes as it is very difficult and costly to bring construction materials from the main land.

Table 1: Area estimates (km^2) of coral reefs in India (DOD-SAC, 1997)

Category	Tamil Nadu	Gujarat	Lakshadweep islands	Andaman and Nicobar islands
Reef flat	64.9	148.4	136.5	795.7
Sand over reef	12.0	11.8	7.3	73.3
Reef vegetation including algae	13.7	112.1	0.4	8.9
Total	90.6	272.3	144.2	877.9

TSUNAMI DAMAGES TO THE CORAL REEFS OF INDIA

Andaman and Nicobar

It has been reported that over 30m high killer wave attacked the Indira point of the Great Nicobar island (Andrews and Vaughan 2005a), the first victim of the Indian land mass, which destroyed the coral reefs, mangroves and beaches of this region. Damage caused by this killer wave was by two actions i.e. the gigantic wave that gushed in to the shallow coastal areas caused substantial physical damage by uprooting the coral colonies and the same wave when receding with almost same power had brought the fallen trees and other solid cargo materials which also caused comparably more physical damage to the coral colonies. It has been reported that several large colonies of over 2 meters diameter have been uprooted and scattered along the shallow region of this cluster of islands. Besides this, large quantities of debris and sediments deposited on the coral remnants may cause long term impacts over the very existence of the coral reefs and the rejuvenation of the affected coral species.

Apart from the physical damage, the tectonic subduction of the earth might have also brought out considerable long-term impact on the coral reefs of this region. Kulkarni (2005) has reported that the coral reefs of Andaman and Nicobar region have suffered heavily by the earthquake by two means, viz: subsidence of southern group of islands and the upliftment of northern group of islands.

Subsided Southern Groups of Islands

Areas of southern group of Andaman - Nicobar islands have been submerged by 1 to 4 meters, while the northern group of Andaman islands have been uplifted due to the tectonic subduction. The Indira point light house in the Great Nicobar island which was nearly half a kilometer away from the shore is now standing alone in the sea (Fig.1a & b). This is one of the examples of subduction of the earth and inundation of seawater on to the land (Kannan 2005*)*. Submergence of shallow coastal areas by 1 - 4 meters in the southern parts will affect some of the coral species which are very sensitive to light. There are also observations that the coral reefs of this region have been suffering due to solar radiation during low tides and hence this subduction of the earth will submerge the corals to safer depths and will help the corals escape from the solar radiation. But there are questions about the survival of light sensitive species and above all the impact of sedimentation on these corals species, as this would favour the settlement of microbial species and cause diseases to the corals. It is also important to note that the tsunami waves while receding could have brought large quantities of microbial load from the land sources and deposited in the marine environment. It has been reported that all the microbial populations have increased in the marine environment to many folds immediately after tsunami. Das et al (2005) have reported two fold increases in total heterotrophic bacteria and one fold increase in actinomycetes population in the Parnagipettai coastal waters, while Ramesh et al (2005) have found that total heterotrophic bacterial load was increased over > 10^5 CFU and the fungal population was doubled after the tsunami in the Chennai coastal waters of Tamilnadu. However, there are questions about the adaptability and survival of these microbes in the newer environment. There is also possibility that continuous tidal flushing might have diluted their population density to certain extent by now.

Figure 1a: Exposed coral reefs at the Indira Point of Great Nicobar (see Light house on the shore)

Uplifted Northern Group of Islands

On the other hand, in the uplifted northern group of Andaman islands, all the coral reefs could have been exposed to the sunlight. Andrews and Vaughan (2005b) have reported that the coral reefs of middle and northern Andaman islands especially along the east coast have been exposed to 0.75 - 1 m height during low tides and are dead. This would further intensify the coral damage by extending the period of exposure and exposure of new colonies. Kulkarni (2005) has stated that the coral reefs of Andaman and Nicobar are dominated by Porites followed by Acropora and any substantial damage to these corals would mean that the entire reef of this region will become unstable and will not be able to withstand any further environmental stress.

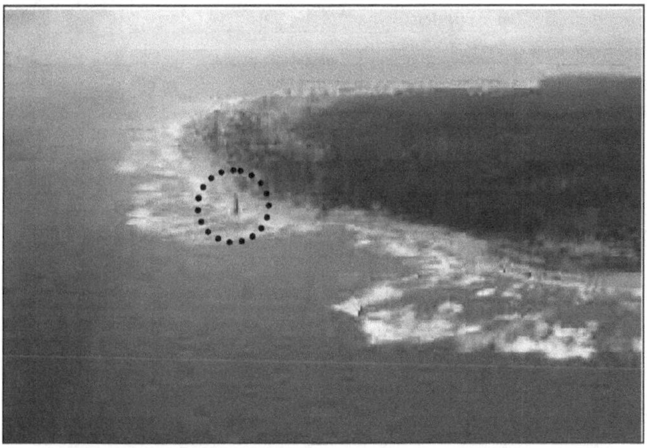

Figure 1b: Submerged Indira point and standing alone light house in the sea.

Estimates of Loss

Regarding the area estimates of coral damages, different figures have been estimated, but all of them agree that at least half of the coral reefs in the Nicobar and southern group of Andaman islands have been damaged seriously. Ramachandran et al (2005) found that over 41% of Camorta, 49% of Katchal, 53% of Nancowry and 59% of Trinkat island coral reefs have been affected. Table-2 shows the estimates of coral damages that have occurred along the Andaman and Nicobar group of islands. But one can expect that there are possibilities of recovery of coral reefs in due course from the sedimentation impacts. Repeated tidal flushings could have washed away certain amounts of the sediments and debris deposited on these reefs, by now. Though the corals are slow growers, there are possibilities of recovery. But the corals which have totally lost their underlying reef structures by the tsunami may not recover. But, in general, only the branching and tabuloid corals have been damaged considerably without much affecting the underlying reef structures. Hence, accurate estimates can be given only after a year of tsunami using high resolution satellite data coupled with extensive field surveys, so as to undertake concrete recovery measures.

Table 2: Coral reef areas (in ha) affected by tsunami in the Andaman Nicobar islands (SAC, 2005)

Island group	Reef type	Pre-tsunami	Post-tsunami	Change
North Andaman	Reef flat	9147	694	- 8453
	Sand over reef	197	2195	+ 1998
	Mud over reef	0	631	+ 631
	Detritus	0	883	+ 883
Middle Andaman	Reef flat	16734	4930	- 11795
	Sand over reef	0	1286	+ 1286
	Mud over reef	-	-	-
	Detritus	0	174	+ 174
South Andaman	Reef flat	9873	4910	- 4963
	Sand over reef	4403	4550	+ 147
	Mud over reef	-	-	-
	Detritus	0	1211	+ 1211
Nicobar	Reef flat	22372	1008	- 21364
	Sand over reef	1120	5233	+ 4759
	Mud over reef	-	-	-
	Detritus	0	1381	+ 1381

Gulf of Mannar

In general, there was no significant damage to the coral reefs of the Gulf of Mannar region as major part of this region fell in the tsunami shadow zone of Sri Lanka. Hence, all the marine resources remained stable without any drastic damage. However, the part of the Bay of Bengal in the Tamilnadu state experienced slightly stronger wave action than the usual ones. This higher wave action has caused sea erosion in some of the Gulf of Mannar islands and broken few coral colonies. Especially the tabuloid coral (Acropora cytherea) and several other delicate coral species have been broken to some extent (Patterson 2005). The Mandapam group of islands, witnessed not much physical damages to the coral reefs or islands (Fig.2 a & b). IRS LISS III remote sensing pictures also clearly indicate that there are no marked variations. However, field visits undertaken immediately after tsunami have confirmed wide spread sediment deposition (over 1 cm) on most of the coral colonies, but further field surveys after 3 months along the Mandapam group of islands have revealed that most of the sediments have been flushed out. The good thing that has happened in the coral reefs of this region is washing out of maximum seaweed biomass by the higher wave action, as this part of the coral reef was suffering a lot by the shadow effects of seaweeds (Thangaradjou et al 2005). Patterson (2005) also found higher amount of washed ashore seaweed and seagrass fragments along the islands and mainland coasts.

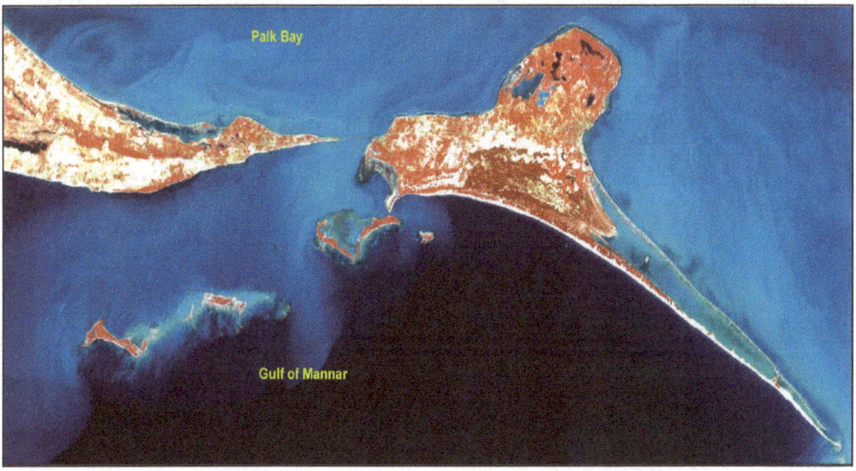

Figure 2a: IRS LISS III imagery of 12th January 2004 showing Mandapam group of islands

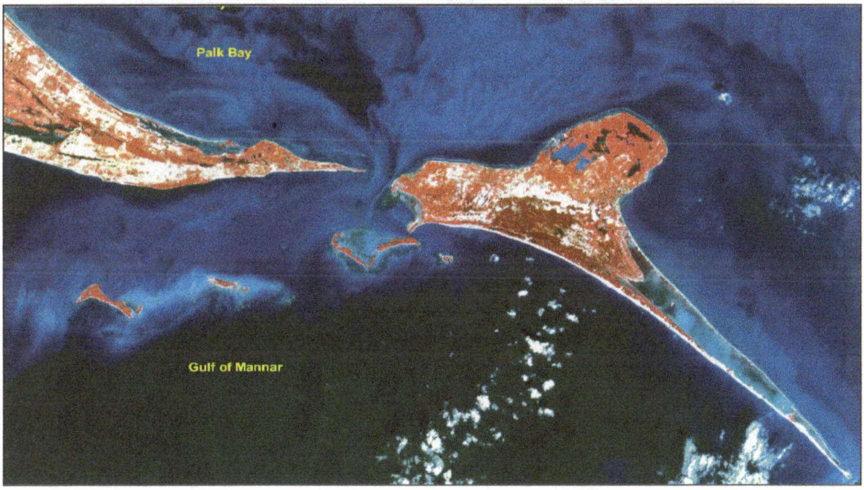

Figure 2b: IRS LISS III imagery of 6th January 2005 showing Mandapam group of islands

Palk Bay

In the case of Palk Bay, satellite remote sensing studies reveal that this part of the Bay of Bengal has not been affected by the tsunami waves, by virtue of being in the "shadow" of Sri Lanka. Coral resources of this part of the Bay of Bengal are still in good condition and even small portions are not lost due to tsunami (Fig.3a & b and Table-3). However, the expected overall changes in the physico-chemical and biological features of the Bay including sedimentation, nutrient enhancement and reduction in seawater temperature may pose problems to the coral resources of this region in future. Kannan et al (2005a) have found reduction in reef areas coupled with increase in reef vegetation (seaweed growth) due to gradual increase in nutrient level in the water column.

Table 3: Bioresources of Palk Bay (pre and post-tsunami)

Sl. No.	Resource category	Area (ha)		
		2004	2005	Difference between 2004 – 2005
1	Reef area	286.95	285.92	(-) 1.03
2	Reef vegetation	155.85	157.80	(+) 2.05
3	Sand over reef	226.53	228.30	(+) 2.23
4	Seagrass bed	123.98	134.66	(+) 11.32
5	Sandy beach	212.23	212.00	(-) 0.23

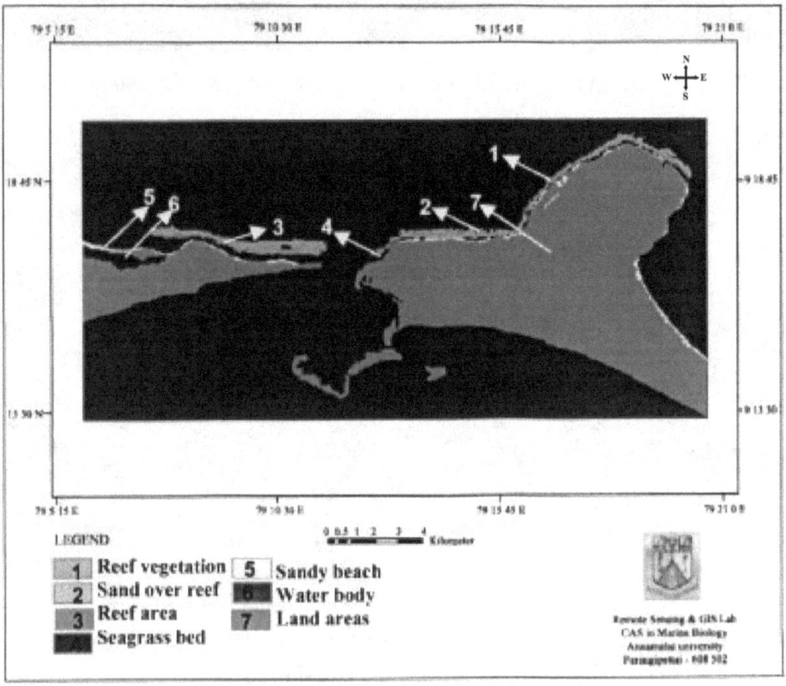

Figure 3a: Resource map of Palk Bay interpreted from IRS LISS III imagery of 12th January 2004

Reflection of coral reef damage

Due to tsunami, over 591 fishing villages and 362 fish landing centres along Tamilnadu coast and 30 out of 38 inhabited islands of Andaman and Nicobar islands were affected badly. The fishery sector has suffered major damage in terms of lives,

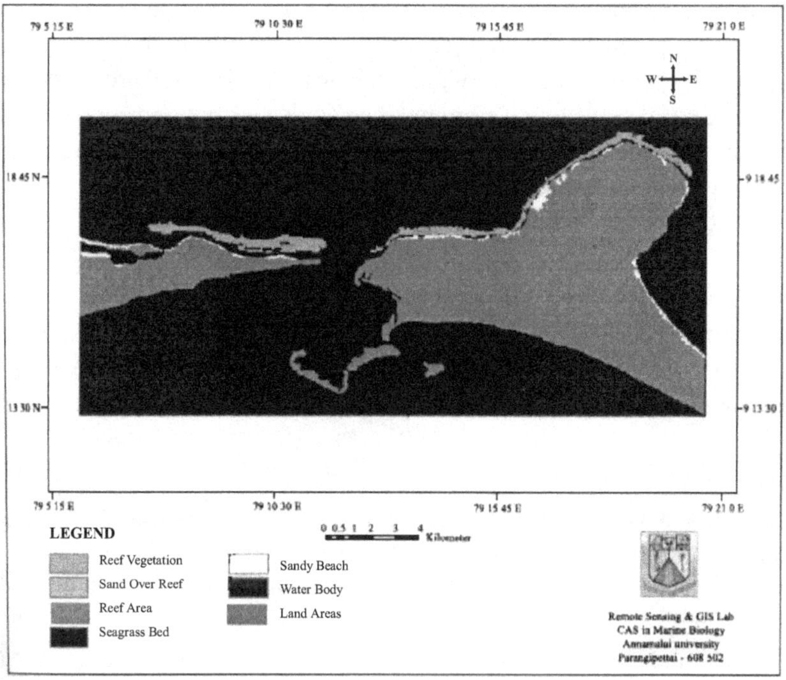

Figure 3b: Resource map of Palk Bay interpreted from IRS LISS III imagery of 6th January 2005

boats, gears and the infrastructure such as *harbours* and fish landing centres. *All these losses* would have affected the coastal human population much. Though many national and international funding agencies including State and Central Governments are supporting them in all possible ways, it will be difficult for them to reach the normal situation in terms of fisheries production. Many fishermen are psychologically not ready to fish again.

Any damage to the coral reefs on a large scale would definitely affect the coastal fisheries. Though the coral reefs are distributed in 1% of the total ocean, it is a proven fact that over 25% of the world fishery production is from the coral reef areas. Due to the damage of coral reefs in Andaman and Nicobar regions, coupled with loss of extensive seagrass meadows in the Gulf of Mannar, there will be reduction in fish catch, because of the fact that the coral reef and seagrass ecosystem are the important nursery and shelter grounds for a plethora of juveniles of commercially important shell and finfish. More importantly, the coral reefs may take a minimum of 10 years to recover to certain extent while the seagrasses will take a year or two to recover. All these would mean that the coastal fishery may face a setback during the coming years, which in turn can adversely affect the socio-economic conditions of the coastal population.

Another important aspect to be considered is the setback in the coastal tourism which is mainly based on the coral reef resources especially in the Andaman and Nicobar, as many people are afraid of undertaking tours to coastal areas after tsunami. Moreover, the coral reefs which attracted the people also got damaged considerably and they may not attract the people much now. Naturalists are happy with this trend as all these coral reefs are now free from the clutches of tourism pressure and they can rejuvenate without much disturbance.

ACKNOWLEDGEMENT

Authors are thankful to the Director, CAS in Marine Biology and authorities of Annamalai University for providing with facilities. Authors also thank the Ministry of Environment and Forests, Government of India for financial support.

REFERENCES

Andrews, V.H. and Vaughan, A. (2005a) Ecological impact assessment in the Andaman islands including observation in the Nicobar islands - Post 2004 tsunami. Report 2005. The Wildlife Trust of India, New Delhi and Madras Crocodile Bank Trust, Tamil Nadu.

Andrews, V.H. and Vaughan, A. (2005b) Ecological impact assessment in the Andaman islands including Little Andaman island after the 2004 tsunami. Interim Report, Andaman and Nicobar Department of Environment and Forests, Port Blair.

Das, S., P.S. Lyla and Ajmalkhan, S. (2005) Variations in microbial population in a mangrove ecosystem in Parangipettai coast (south east coast of India): Effect of tsunami?. Research Journal of Chemistry and Environment, v. 9(3), pp. 46-49.

DOD and SAC (1997) Coral reef maps of India. Department of Ocean Development and Space Application Centre, Ahmedabad, India.

Kannan, L. (2005) Rapid assessment on the Impact of tsunami: destruction of mangroves along the Nicobar islands and corals of the Palk Bay. Rapid Assessment Report, Ministry of Environment and Forests, New Delhi and CAS in Marine Biology, Parangipettai.

Kannan, L., Thangaradjou, T. and Duraisamy, A. (2005a) Effects of tsunami on the coastal and marine biodiversity of India. In: Tsunami: the Indian Context, edited by S.M. Ramasamy and C.J. Kumanan, Bharathidasan University, Tiruchirapalli, pp. 261 – 266.

Kannan, L., Thangradjou, T., Sridhar, R. and Senthilkumar, S. (2005b) Evaluation of the Palk Bay bioresources through conjunctive use of conventional ground surveys and satellite remote sensing. Final Report, Ministry of Environment and Forests, Government of India and CAS in Marine Biology, Parangipettai.

Kulkarni, S. (2005) Tsunami impact assessment of coral reefs in Andaman and Nicoabr islands. Reef Watrch Marine Conservation, Mumbai, India.

Ming, C.L. (1997) Southeast Asia as the global centre of marine biodiversity. Tropical Coasts, v. 4(1), pp. 4-8.

Patterson, J.K. (2005) Rapid assessment of status of corals in the Gulf of Mannar after tsunami. Assessment Report, Suganthi Devadason Marine Research Institute and Tuticorin Port Trust, Tuticorin.

Ramachandran, S., Anitha, S., Balamurugan, V., Dharanirajan, K., Ezhil Vendhan, K., Marie Irence Preeti Divien, Senthil Vel, Sujjahad Hussain, I. and Udayaraj, A. (2005) Ecological impact of tsunami on Nicobar Islands (Camorta, Kathcal, Nancowry and Trinkat). Current Science, v. 89(1), pp. 195 - 200.

Ramesh, S., Jayaprakashvel, M. and Mathivanan, N. (2005) Microbial diversity in Bay of Bengal of Chennai coasts in pre and post-tsunami periods. In: Proceedings of the National Seminar on effect of tsunami on coastal ecosystem, CAS in Marine Biology, Parangipettai, pp. 40 - 42.

SAC (2005) Assessment of damages to coastal ecosystems due to the recent tsunami. Summary Report, Space Application Centre, Ahmadabad.

Thangaradjou, T., Kannan, L. and Balasubramanian, R. (2005) Seaweed harvesting as coral reef management practice – A case study on coral reefs of the Gulf of Mannar Biosphere Reserve. In: Proceedings of the National Seminar on Rejuvenation and reclamation of coral reefs in Gulf of Mannar, Kamaraj College, Tuticorin, pp. 71 – 83.

Venkataraman, K. (2003) Coral reefs of India. In: Natural Aquatic Ecosystem of India, Thematic Biodiversity Strategy and Action Plan, The National Biodiversity Strategy Action Plan, Zoological Survey of India, India, pp. 115-136.

Wilkinson, C. (2002) Status of coral reefs of the world. Global Reef Monitoring Network and Australian Institute of Marine Science, 378 p.

GEOMATICS IN TSUNAMI
S.M. Ramasamy, C.J. Kumanan, Brig. R. Sivakumar and Bhoop Singh (eds.), pp. 179-190
New India Publishing Agency, New Delhi (India)
E-mail : spjain_niph@rediffmail.com Web: www.bookfactoryindia.com

14

REFLECTIONS OF RECENT TSUNAMI ON MANGROVE ECOSYSTEM, CUDDALORE COAST, INDIA

V.Selvam

Abstract

The mangrove wetlands form dynamic ecotones between land and sea. Mangrove wetlands are unique in terms of environment, plant and animal communities and their distribution is restricted in tropical and subtropical coastlines. They perform a number of protective and productive functions, which ensures ecological and livelihood security to the coastal communities. The role of mangroves in mitigating the impact of cyclones and storm surges is well known but only limited information is available on their role in reducing the impact of tsunami. The Pichavaram mangroves, located on the southeast coast of India in Tamil Nadu, which was hit by the 26 December 2004 tsunami, provided opportunities to understand the contribution of it in tsunami mitigation. Through a systematic field study, a comparative analysis was made on the impact of tsunami in a set of villages, one was protected by mangroves and the other lacks any vegetation surrounding it. These villages are located at more or less equal distance from the sea and situated at about same elevation. Assessment carried out with well defined indicators such as loss of lives due to forceful entry of tsunami water, damages to houses and types of houses, etc., shows that loss of lives and damage to houses were much less in the village surrounded by Pichavaram mangroves. Thus, Pichavaram mangrove wetland, which was restored with the participation of the local community, Tamil Nadu Forest Department and M.S.Swaminathan Research Foundation (MSSRF) during 1993-2004 provided great dividend during tsunami. This underlines the importance of strengthening the ecological foundation of the coastal zone with mangroves and other vegetation as one of the strategies for mitigating the impact of natural disasters in the coastal areas.

Keywords: Mangroves, Natural Disasters, Tsunami, Impact Reduction and Pichavaram.

INTRODUCTION

Coastal zones, throughout the world, are endowed with a variety of ecosystems, which perform a number of protective and productive functions to ensure ecological security against natural calamities such as cyclones, storm surges and tsunami and livelihood security of the coastal community. Mangrove wetland is one such important ecosystem of the tropical and subtropical coastlines. Mangroves are defined as assemblages of trees and shrubs that grow in the areas where river water mixes with seawater. Mangrove ecosystems have two components, mangrove forest and associated water bodies. A group of woody trees and shrubs that grow well in saline water and water logged areas constitute the forest component. Whereas the tidal channels and canals that intersect mangrove forest and fragment them into small islands, shallow lagoons and bays found associated with the mangroves constitute water bodies. The mangrove forest and associated water bodies are together constitute mangrove wetlands.

Mangrove wetlands are multi - benefit ecosystems that provide protective, productive and economic benefits to local communities. The economic value of the mangroves are the availability of wood products from the mangrove stem, non-wood products such as fodder, honey, wax and tannin and aquatic products such as fish, prawn, crabs etc. In addition, mangroves provide a variety of other benefits: i) act as a barrier against cyclonic storms and tsunami and prevent the entry of saline water into the land, ii) act as a buffer against floods and thereby reduce coastal erosion, iii) act as breeding, spawning, hatching and nursing grounds for a variety of commercially important fish, prawn and crabs, iv) enhance fishery production of the adjacent neritic waters by exporting nutrients and detritus and v) provide habitats for wildlife ranging from migratory birds to estuarine crocodiles (Blasco et al 1996 and Nguyen et al 1998).

The role of mangroves as bioshield against cyclones and storms are well known since many decades by local community, but only the tsunami 26 December 2004 provided opportunity to understand the protective role of mangroves against tsunami. The present paper provides an overview of the mangroves of India in general and Tamil Nadu in particular and discusses the efforts taken to restore mangrove wetlands of Tamil Nadu and how such restored Pichavaram mangroves played a vital role in mitigating the impact of the recent tsunami on lives, properties and livelihood of the adjoining villages.

MANGROVE WETLANDS OF INDIA

The east and west coasts of the mainland of India and Andaman and Nicobar Islands are characterized by the presence of mangrove wetlands. On the macro scale, geomorphic settings of the mangrove wetlands of the east coast of India are different from that of the west coast. As the western slope of the western ghats are very steep, the rivers directly gush into the Arabian sea without having their actual life cycles and intern developing any characteristic landforms like flood plains, deltas, estuaries etc. Hence, the fluvio - marine systems which only great avenues for mangrove ecosystems are also very thin. As a result, mangrove wetlands of the west coast of India are scanty, small in size and less complicated in terms of tidal creek network. On the other hand, mangrove wetlands of the east coast are larger in area and water bodies associated with mangroves are characterized by the presence of larger brackish water bodies and complex network of tidal creeks and canals (Selvam 2003). This is due to the low easterly sloping topography and the resultant development of well evolved river/streams and fluvio – marine geomorphic systems (Table-1).

Table 1: Mangrove wetlands of India

State	Mangrove wetland	Total area of the wetland (ha)*	Actual forest cover (ha)
East Coast West Bengal	Sunderbans	4,26,000	2,12,500
Orissa	Mahanadi	67,000	21,500
Andhra Pradesh	Godavari	33,250	24,100
	Krishna	25,000	15,600
Tamil Nadu	Pichavaram	1,300	900
	Muthupet	13,000	1,200
West Coast Gujarat	Gulf of Kutchchh	58,200	85,400
	Gulf of Khambhat	53,123	17,700
Other mangroves	-	-	11,600
Andaman and Nicobar islands	Andaman islands Nicobar islands	-	92,900 3,700
Total			**4,87,100**

* Records of the State Forest Department

However the health of the mangrove wetlands with reference to hydrological and soil conditions and wealth of the mangrove wetlands in terms of aerial coverage, species diversity, biomass and productivity are determined by i) degree of protection against high-waves, ii) quantity and duration of freshwater flow and sediment supply and iii) larger tidal amplitude and gently sloping coastal topography. Most of the mangrove plants require low salinity condition for their growth and reproduction (Smith 1992). Hence, luxuriant mangrove forests can be seen only in the estuarine regions where large amount of freshwater is discharged for longer period of time in a year. The area of the mangrove wetlands is determined by the tidal amplitude (difference between high tide and low tide) and slope of the coastline (Duke 1992).

According to Selvam (2003) there is a gradual decrease in the quantity and periodicity of freshwater discharge into the mangrove wetlands as well as tidal amplitude from the northernmost mangroves of Sunderbans of West Bengal to southernmost mangroves of Pichavaram and Muthupet of Tamil Nadu. As a result of these, the aerial extent, species diversity, structure and biomass of the mangrove forest differ widely along the east coast of India. According to Forest Survey of India, out of 4, 87,100 ha of the mangrove wetlands of India, nearly 56.7% (2, 75,800 ha) is present along the east coast and 23.5% (1, 14,700 ha) along the west coast and the remaining 19.8% (96,600 ha) is found in the Andaman and Nicobar islands (Table-1).

MANGROVE WETLANDS OF TAMIL NADU

Tamil Nadu has a coastline of over 1000 km. along the coastline; major mangrove wetlands are present in two areas, one at Pichavaram in Cuddalore District and the other in Muthupet region in Thiruvarur-Thanjavur Districts. Small patches of mangroves are also present along Palk Bay, particularly in Devipattinam region and also in some of the islands of the Gulf of Mannar in Ramanathapuram District. Pichavaram region was hit strongly by the tsunami as it was directly facing the on rushing tsunami waves. Impact of tsunami in Muthupet mangrove region was not strong due to presence of large mud banks and shoals in the Palk Bay and the island of Sri Lanka in the east (Earthquake Engineering Research Institute 2005).

PICHAVARAM MANGROVE WETLAND

Pichavaram mangrove wetland is located in the northern most part of the Cauvery delta. It is situated in between Vellar estuary in the north and Kollidam (Coleroon) estuary in the south and these two are connected by a large brackish water canals. The Pichavaram mangrove wetland consists of 3 Reserve Forests (RF), namely Killai, Pichavaram and Pichavaram Extension area. In 1897 when this mangrove wetland was declared as Reserve Forest, the area was about 1358 ha. According to the remote sensing data of 1986, the total area of these RFs was about 1474 ha. Out of the same, thick mangrove forest occupied an area of 325 ha (22%) and degraded mangrove forest was found in about 375 ha (25%). The area of the water bodies associated with the Pichavaram mangrove wetland was about 380 ha (25.7%). In addition, sandy area, which is not suitable for mangroves due to higher elevation, occupied about 394 ha (26.7%). The above data indicates that nearly 53% of the mangrove forest area (not the wetland area) was in degraded state in 1986.

Pichavaram mangrove is rich in fishery resources. According to Chandrasekaran and Natarajan (1992), a total amount of 237 tones of fish is harvested annually from this mangrove wetland. Pichavaram mangrove is also rich in fodder resources. Fishing and farming communities belonging to 17 hamlets utilize the resources of the Pichavaram mangrove wetland. A total number of 1,900 fishers are annually dependent on the fishery resources for their livelihood; some 1,000 fishers fish seasonally in the mangrove waters (Selvam et al 2003).

RESTORATION OF PICHAVARAM MANGROVE WETLAND

Causes of Degradation

Ecological studies carried out in the Pichavaram wetland by MSSRF between 1993 and 1995 showed that unscientific management practices followed in the past are the main causes of degradation (Selvam et al 2002, 2005). In the Pichavaram mangrove wetland, a management method called "coupe-system" was followed from 1935 to 1970. Under this system, healthy mangrove forest was clear-felled by government agencies in coupes by rotation once in 20 years for revenue generation. This triggered a chain reaction, leading to development of hyper-saline conditions in coupe-felled area, leading to the drastic reduction in the natural regeneration of mangroves.

Since nearly 80% of the volume of the mangrove surface soil is made up of water, exposure of this soil to the sun due to clear felling caused evaporation of soil water. This in turn led to subsidence of sediment in the clear felled area, on account of which the topography of the coupe-felled area became trough shaped. As a result, tidal water entering into these "troughs" during high tide became stagnant; evaporation of stagnant tidal water led to increase in salinity, which is lethal to any mangrove plant. An estimate indicates that coupe felling is responsible for nearly 65% of degradation in the Pichavaram mangroves (Selvam et al 2002).

Development and Demonstration of Restoration Technique

Development and demonstration of a restoration technique began late 1995 by MSSRF in the Pichavaram mangrove wetlands. A clear-felled area of 8 ha, where topography had become trough-shaped, was selected to demonstrate the restoration technique. This technique is simple. The trough-shaped area was connected to a natural canal nearby, through a long and deep artificial canal from which a number of feeder canals were dug to cover the entire degraded area. This enabled tidal water to freely flow in and out of the degraded area (Fig.1). As a result, salinity of the degraded area reduced drastically with increase in soil moisture. Propagules *of Rhizophora* sp and *Avicennia marina* seedlings were planted along main and feeder canals respectively at a distance of 1 x 1 m. A total number of about 80,000 seedlings were planted in the demonstration site during December 1995 and now nearly 80% of the saplings have grown into tall trees. At the end of the demonstration of restoration techniques, three important questions cropped up: i) who will maintain the canals if they silt up? ii) How to avoid social pressure, if any? and iii) how to upscale the restoration activities?

Figure 1: Canal system developed jointly by MSSRF, Forest Department and local community is very effective in restoring the degraded mangroves of Tamil Nadu

Joint Mangrove Management in Tamil Nadu

Joint Mangrove Management (JMM), which is defined as a process by which all the stakeholders jointly plan, implement and monitor mangrove restoration and conservation activities through participatory analysis, was introduced as an answer to the above questions. It was introduced in Pichavaram and then to Muthupet mangrove wetlands and finally to other mangrove wetlands of Andhra Pradesh, Orissa and West Bengal. The restoration and conservation activities were undertaken with a tripartite agreement amongst the local community, concerned state Forest Department and MSSRF. This programme was implemented in eight hamlets of Tamil Nadu (4 in Pichavaram and 4 in Muthupet) till May 2003, covering traditional and non-traditional fishing and farming communities. The following are the major achievements of JMM in Tamil Nadu.

- A science-based, community-centered and process oriented approach was developed and demonstrated to restore, conserve and sustain mangrove wetlands.
- Eight village-level institutions have been formed with 885 families as members to plan and implement JMM and socio-economic development programmes.
- A total area of 675 ha has been restored and healthy mangroves in 2,720 ha are being protected by the above village-level institutions. The entire degraded areas of Pichavaram have been restored (Fig.2b).
- A total number of 5.5 million saplings (4.8 million of *A. marina* and 0.7 million of other species) have been planted by the local community; average survival is 68%.
- A total number of 85 self-help groups (50 of women and 35 of men) have been formed with 815 members belonging to the poor and the poorest sections

of the mangrove-dependent community. 16 types of micro-enterprises, both group-based and individual-based, covering 402 families have been initiated.

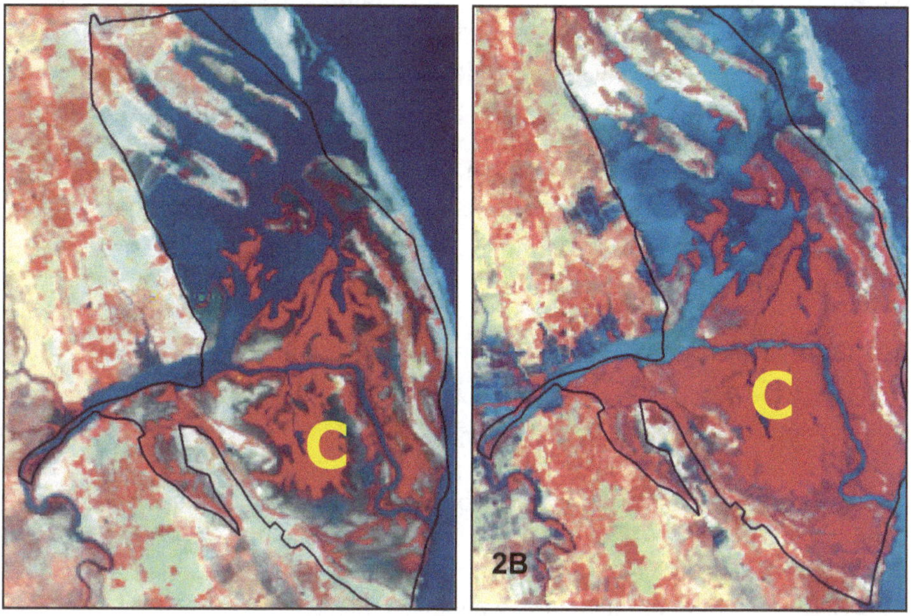

Figure 2: Remote sensing imagery of Pichavaram mangroves. 2A - before restoration (Landsat 5 TM, 1986) and 2B – after restoration (IRS ID LISS III, 2002). Black line indicates Reserve Forest boundary and C - indicates mangroves. (Note the drastic increase in grey tone area indicating the prolific growth of mangroves(C) in between both periods).

The Ministry of Environment and Forests (MoEF) Government of India formed a sub-committee which examined the JMM models implemented in Tamil Nadu, Andhra Pradesh and Orissa by MSSRF and observed that this was the best available model. The MoEF has now included this JMM model as one of the strategies for conservation and sustainable management of mangrove wetlands envisaged in its National Mangrove Action Plan.

PICHAVARAM MANGROVES AND TSUNAMI

Local communities living around mangroves and other coastal tree vegetation are aware of the effectiveness of these vegetations as shields against cyclones, hurricanes, storms and tsunami. For example, people in the Muthupet region of Tamil Nadu have been calling mangrove forest as "*aalayaathi kaadu*" for thousands of years. "*Aalai*" means waves and "*aathi*" means mitigate and "*kaadu*" means forest. This traditional wisdom is supported by experimental studies in the field and laboratory. Quantitative studies in Vietnam show that thick and tall *Kandelia* mangrove trees in a 100 m belt would reduce the impact of wind induced waves by as much as 20% (Mazda et al 1997). Similarly, a series of hydraulics experiments and numerical simulation studies conducted in the laboratory in Japan show that 30 trees per 100^2 in a 100 m wide mangrove belt would reduce elevation and velocity of tsunami waves by 90% (Hiraishi 2002). However, field based evidences that provide information for planners and policy makers to support mangrove restoration and green belt development along the coastal areas, as a measure to reduce the impact of tsunami and other coastal events such as hurricanes and storms, is very limited (Danielsen et al 2005). The Pichavaram mangroves present in Cuddalore coastal zone, which was seriously affected by the 26 December 2004 tsunami, provided opportunities for evaluating the functions of mangroves towards tsunami mitigation.

Rapid Assessment

During the first week of January 2005, a rapid assessment was undertaken to study the impact of tsunami on lives and property of the 11 villages that are located around Pichavaram mangroves. Impact on mangroves was also assessed. In the case of mangroves, information was collected on kind of damages occurred to mangrove trees and the role they played in mitigating the impact of tsunami. In the villages, the data on loss of lives, loss of fishing crafts and gears, damage to houses and damage to agriculture land etc., were also collected.

Scientific Assessment

As per the rapid assessment, the impact of tsunami was very severe in the villages that are without any mangrove forest around them and was less in the villages that are surrounded by mangroves. But further analysis raised the question that whether less impact in the villages were due to the distance at which they are located or because of the protection provided by the mangroves. This doubt was raised because all the severely affected villages were located close to the sea whereas less affected villages were located about 1 to 3 km. away from the shore. Hence, the following criteria were used for the selection of villages for scientific assessment:

- Villages should be located within a distance of 2.0 km from the shore, a distance upto which tsunami water penetrated (Chandrasekaran et al 2005),
- Mangrove protected and unprotected villages should be located almost at same distances from the coast and on the same elevation,
- Mangrove protected and unprotected villages should be more or less similar in economic status and community should be willing to participate in the assessment.

Out of 11 villages, only two villages namely, Muzhukkuthurai and T.S. Pettai met these criteria. Muzhukkuthurai is located in the northern side of the Pichavaram mangrove wetlands and the distance between the village and the sea is about 1400 m. No mangrove forest is present between the village and the sea. T.S.Pettai is located about 1600 m from the sea and thick mangroves of about 80 ha present between the village and the sea.

Indicators for Assessment

Impact of tsunami on the villages is relating to loss of lives, damages to houses and household properties, loss of boats and nets, inundation of saline water into agriculture lands and water bodies, etc. Some of the losses, for example, loss of boats and nets mostly happened in the boatyards (either in a jetty as in the case of larger villages or along the beaches) where they were kept and neither mangroves nor shelter plantations cover these boatyards. Hence, loss of boats and nets common to both protected and non-protected villages and thus, could not be taken as an indicator to assess how mangroves lessened the impact of the tsunami. Total loss of human lives in protected and non-protected villages also could not be taken as indicator because some of them died while near the sea and others when they were in the village. All these issues were discussed with the villagers. On the basis of the above discussions the following three indicators were selected to understand how mangroves mitigated the impact of the tsunami on the lives and properties of the villages: i) loss of lives within the village due to forceful entry of the tsunami waves (loss of lives nearby the sea excluded), ii) number and type of houses damaged (both fully and partially) and iii) height of the water entered into the villages (identified as lines in the walls of the houses).

How Mangroves Reduced the Impact of Tsunami

Loss of lives, damages to houses along with other details pertaining to Muzhukkuthurai and T.S.Pettai villages are given in Table-2. As indicated in the table, loss of lives and damage to houses was less in the mangrove protected T.S.Pettai village than in the non-protected villages. Field visits and observations of the local community during tsunami indicate that the mangrove forest that lies between the sea and T.S.Pettai village played an important role in mitigating the impact of the tsunami. This mangrove forest has three distinct vegetation zones, namely *Rhizophora*, *Avicennia* and *Excoecaria* zones. The *Rhizophora* zone, which is about 10 m wide, is located along the edge of the lagoon and is characterized by the presence of mainly *Rhizophora apiculata* and *Rhizophora mucronata* trees, which possess strong stilt roots that give additional support to the tree. Next to this *Rhizophora* zone, *Avicennia* is present, which is characterized by the presence of mainly *Avicennia marina* trees. The breadth of this zone is about 900 m. Near to the village, sandy area of about 10 to 12 m in breadth is present, which is characterized by the presence of another mangrove species *Excoecaria agallocha*. Field visits showed that tsunami waves, after crossing the lagoon by breaching the sandy beach, first traversed through the *Rhizophora* zone and then through *Avicennia* and *Excoecaria* zones and finally reached T.S.Pettai village. In this process, *Rhizophora* trees along the edges with their distinct stilts roots absorbed most of the energy of the tsunami waves, which is indicated by the presence of a few uprooted trees and broken branches in the *Rhizophora* zone. The speed of the waves was reduced and their destructive energy was lost as they traversed through *Avicennia* and *Excoecaria* zones. Apart from this, most of the water brought in by the tsunami was distributed into the lagoon, canals and tidal creeks associated with the mangrove and hence, the volume of water that reached the village was less, as also narrated by the villagers, is the yet another reason for reduced impact in T.S.Pettai and other villages that are located within the physical cover of the mangroves.

Table 2: Impact of tsunami in Muzhukkuthurai (mangrove non protected village) and T.S.Pettai (mangrove protected village) of Pichavaram region

Village	Distance from the sea	Elevation	Area of mangroves	Loss of lives			Damage to property		
				Population	Death	%	Total houses	Damaged	%
Muzhukk-uthurai	1400 m	1.8 m	0 ha	545	11	2	153	50	98
T.S.Pettai	1600 m	2.2 m	80 ha	1125	0	0	225	0	0

In the case of Muzhukkuthurai village, the lagoon between the village and the sea, as explained by the local people, was able to absorb considerable amount of water brought in by the tsunami but the quantity of water and speed with which tsunami wave entered into the village was high. The tsunami waves completely pulled down almost all the brick walled huts and caused extensive damages even to concrete houses.

CONCLUSIONS

In areas of maximum intensity nothing would have prevented disastrous effect of tsunami. But, as indicated in the Pichavaram region, the coastal vegetation such as mangroves can play an important role in mitigating the impact of tsunami. During the Orissa Super Cyclone of 1999 it was proved that mangroves were effective in reducing the impact of cyclones. As explained by Dr.M.S.Swaminathan, Chairman, National Commission on Farmers, Government of India, coastal plantations including mangroves should become part of strengthening the ecological foundation of the coastal areas (Swaminathan 2005). He said that coastal shelterbelt plantations, apart from acting as bioshields, also serve as carbon sinks, since they help to enhance carbon sequestration and thereby contribute to reduce the growing imbalance between carbon emissions and absorption. Hence, restoration of degraded mangroves and conservation of existing healthy mangroves should be given priority in the disaster preparedness strategy for the coastal areas. Apart from the existing mangroves, there are opportunities to create mangroves in newer areas. Site selection for mangrove generation need proper studies as mangroves grow in specialized environmental conditions.

REFERENCES

Blasco, F., Saenger, P. and Janodet, E. (1996) Mangroves as indicators of coastal change. Catena, v. 27, pp.167-178.

Chandrasekaran, V.S. and Natarajan, R. (1992) Small scale fishery of Pichavaram mangrove swamp, southeast cost of India. Naga, v.15(2), pp. 41-43.

Chandrasekaran, H., Singh, V.P., Rao, D.U.M., Nagarajan, M. and Chandarasekharan, B. (2005) Effect of tsunami on coastal crop husbandry in parts of Nagapattinam district, Tamil Nadu. Current Science, v. 89(1), pp. 30-32.

Danielsen, F., Sorensen, M.K., Olwig, M.F., Selvam, V., Parish, F., Burgess, N.D., Hirashi, T., Karunagaran, V.M., Rasmussen, M.S., Hansen, L.B., Quarto, A. and Suryadiputra, N. (2005) The Asian tsunami: a protective role for coastal vegetation. Science, v.310, 643p.

Duke, N.C. (1992) Mangrove Floristics and Biogeography. In: Tropical Mangrove Ecosystem, edited by Robertson, A.I and Alongi D.M., Americal Geophysical Union, Washington DC, pp. 63-100.

Earthquake Engineering Research Institute (2005) The Great Sumatra Earthquake and Indian Ocean tsunami of December 26, 2004. Special Earthquake Report No.3, Earthquake Engineering Research Institute, USA, pp. 1-12.

Forest Survey of India (1999) State of the Forest Report. Forest Survey of India, Dehradun, pp. 32-43.

Hiraishi, T. (2002) Tsunami Risk and Countermeasure in Asia and Pacific Area: Applicability of Greenbelt Tsunami Prevention in the Asia and Pacific Region. Institute of Earthquake and Tsunami Mitigation in Asia Pacific, Japan, pp. 1-6.

Mazda, Y., Magi, M., Kogo, M. and Hong, P.N. (1997) Mangroves as a coastal protection from waves in the Tong King delta, Vietnam. Mang. Salt Marsh, v. 1(2), pp. 127-135.

Nguyen, H.T., Adger, W.N. and KELLY, P.M. (1998) Natural resource management in mitigating climate impacts: the example of mangrove restoration in Vietnam. Global Environmental Change, v. 8(1), pp. 49-61.

Selvam, V. (2003) Environmental classification of mangrove wetlands of India. Current Science, v. 84(6), pp. 757-765.

Selvam, V., Gnanapazham, L., Navamuniyammal, M., Ravichandaran, K.K. and Karunagaran, V.M. (2002) Atlas of Mangrove Wetlands of India. Part I TamilNadu. M.S.Swaminathan Research Foundation, Chennai, India, pp. 1-99.

Selvam, V., Ravichandran, K.K., Karunagaran, V.M., Mani, K.G., Evanujaline, G.J.B. (2003) Joint Mangrove Management in Tamil Nadu: Process, experiences and prospects, Part 1 Situation Analysis. M.S.Swaminathan Research Foundation, Chennai, pp. 27-38.

Selvam, V., Ravishankar, T., Karunagaran, V.M., Ramasubramanian, R., Eganathan, P and Parida, A.K. (2005) Tool kit for establishing coastal bioshield. M.S.Swaminathan Research Foundation, Chennai, India, pp. 44-52.

Smith, T.J. (1992) Forest Structure. In: Tropical Mangrove Ecosystem, edited by Robertson, A.I and Alongi D.M., American Geophysical Union, Washington DC, pp. 101-136.

Swaminathan, M.S. (2005) Beyond Tsunami: Saving Lives and Livelihoods. National Commission on Farmers, Government of India, New Delhi, pp. 1-10.

GEOMATICS IN TSUNAMI
S.M. Ramasamy, C.J. Kumanan, Brig. R. Sivakumar and Bhoop Singh (eds.), pp. 191-198
New India Publishing Agency, New Delhi (India)
E-mail : spjain_niph@rediffmail.com Web: www.bookfactoryindia.com

15

TSUNAMI INDUCED WATER QUALITY CHANGES IN BUCKINGHAM CANAL, CHENNAI REGION, INDIA

Usha Natesan and S.Kalaivani

Abstract

Ennore is an industrial region having many industries on chemicals, fertilizers, refineries etc., discharging their effluents into the Buckingham canal. Present study focuses on the impact of tsunami on water quality of Buckingham Canal segment found to the north of Chennai city, Tamil Nadu, India, in addition to the human interferences. Water samples were collected from surface and near bottom at five locations covering about 20km stretch of the Buckingham Canal during high tide on the spring tide day before (December 2004) and after (January 2005) the mega tsunami. Various water quality parameters measured were pH, Temperature, Salinity, Dissolved Oxygen, Suspended Particulate Matter, Nutrients and Chlorophyll. The maximum depth observed in the canal was 2.5m in December, whereas a shallow depth of less than 0.5m was recorded in January. The water quality analyses confirmed the incursion of saline water (upto 32ppt) into the Canal. Not much variations were observed in Dissolved Oxygen and pH. Because of the organic pollution in the canal, the water was rich in nutrients before tsunami, whereas due to the large quantity of seawater invasion, dilution had occurred after tsunami. High turbidity was observed in January 2005 indicating the tsunami induced disturbance and resuspension of the sediments from sea bed.

Keywords: Tsunami, Ennore, Tamil Nadu, Buckingham Canal.

INTRODUCTION

Man has started increasingly depending on the sea for various resources, such as minerals, food, petroleum, gas etc. Discharge of hazardous chemicals,industrial wastes, radioactive materials and various toxic substances into the sea, harms the living resources and disturbs the marine related ecosystems. Different levels of pollution in the coastal waters in general and in estuaries in particular define variations in the carrying capacity of the ecosystems.

Economic ramifications due to marine ecosystem damages were explained in detail in EPA (1981). Boesh (1982) inferred that alterations in nutrients and organic inputs to the marine ecosystem can revoke changes in the species composition. Levin (1983) observed that Laboratory tests must be conducted by keeping the ecosystem and the fragile and sensitive species and their structure & functions in mind. The distribution and abundance of the copepods in the pollution gradient zones of the Bombay Harbour-Thane Creek-Bassein Creek were analysed by Ramaiah and Nair (1997). Evaluation of the level of pollution at Bassein and Malad Creeks through sequential comparsion and saprobic indices was carried out by Praveen Kumar et al (1999). USEPA report stated that the excess of phosphorus and nitrogen present in the Chesapeake Bay in North America had mired the Bay's health to the tune of 20%. Under this back drop, the present study was under taken, whether at all any changes have occurred in water quality of buckingam canal due to the recent tsunami.

Tsunami

A devastating earthquake had occurred on December 26th 2004 off the west coast of Sumatra, Indonesia. This resulted in tsunami which killed thousands of people and caused greater loss to the property in Southern India, Srilanka, Indonesia, etc. In India, Andaman and Nicobar Islands, TamilNadu, Pondicherry, Andhra Pradesh, Kerala were the worst hit.

Study Area

The Buckingham Canal from Ennore creek to the mouth of Pulicat lagoon (located to the north of Chennai City, India) is heavily polluted due to the discharges from industries, municipal sewage, thermal power stations and aquafarms. After the construction of satellite port, the situation has become worse due to the accumulation of pollutants in the canal. This has already started creating hazards to the human health and hindrance to aquatic animals. In addition to this man-made stresses, December 26, 2004 tsunami had changed the environmental conditions of the area unpredictably. From the post-tsunami field investigations, it was understood that waves struck 3 times, with the first one being the biggest with the run-up height of 4.5m. A site visit after tsunami further revealed, the water exchange between the sea and the polluted water in the Canal. So, a study had been taken up on the water quality to assess the environmental status and biodiversity of the Buckingham Canal due to tsunami.

METHODOLOGY

Sampling Strategy

Buckingham Canal is a Zero Canal extending from Kakinada (Andhra Pradesh) in the north to Marakkanam backwaters, near (Pondicherry) in the south along the east coast of India. It is laid approximately 200-1000 meters west of the coastline all along. The proposed study area covers around 20 km stretch of the Buckingham Canal from Pulicat Lagoon in the north to Ennore Creek in the south, located on the northern side of Chennai city, India. In the study area, there are two openings into the sea, of which one at the Ennore Creek is (Fig.1) kept open by continuous dredging by the thermal power plant authorities and the second one is (5, Fig.1) closed permanently by the sandbar formed by littoral currents. Based on the various environmenal stresses, 5 sampling stations were identified from south to north.

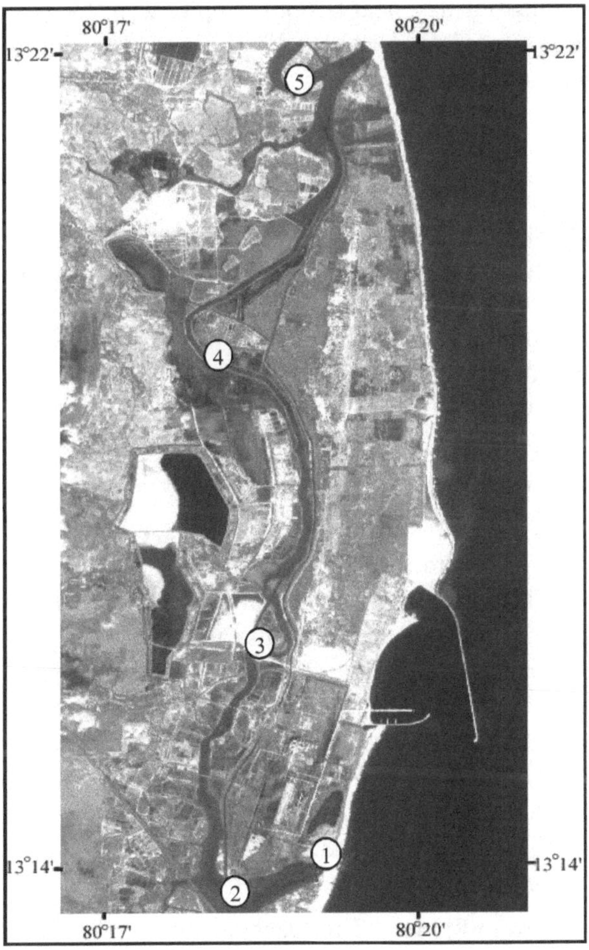

Figure 1: Study Area 1-5 Sampling Locations

- Station 1 - near Ennore creek mouth where coolant water is discharged from the thermal power plant
- Station 2 – Ennore Creek where wastes are discharged by many industries
- Station 3 - in the Buckingham canal, 5 km north of station 2
- Station 4 – in the canal north of station 3 near aquaculture farms and
- Station 5 – in the northern most end near the northern of the canal, close to the Pulicat lagoon

Study area along with the sampling stations is shown in Fig.1 GPS (Global Positioning System) was used in fixing the sample locations.The sampling was carried out by hiring a boat. Water samples at surface and near bottom were collected in December 2004 (pre-tsunami), where as in January 2005 (post-tsunami) during high tide on the spring tide day, near bottom samples were collected using Routner water sampler. The samples were labelled by unique identifiers and stored in ice immediately after collection in the field. They were brought to the laboratory for analysis.

Analysis

Depth of the stations was measured using a graduated pole. The physical parameters like pH and temperature were measured in the field itself. The parameters viz., Dissolved Oxygen (DO), Salinity, Suspended Particulate Matter (SPM), Nutrients and Chlorophyll were analysed in the laboratory following the procedure developed by Grasshoff and Kremlig (1998).

RESULTS AND DISCUSSIONS

Physiochemical Characteristics

The variations in temperature between surface and bottom is negligible indicating the absence of stratification. Though the station 1 recorded over 8°C more than the other stations during pre-tsunami (Dec,2004) and 4°C during post - tsunami (Jan,2005), no marked change was observed in between pre and post-tsunami as far as this station 1is concerned. This leads to the surmise that while the temperature almost remained static in station 1; hence in other four stations it might have increased (Fig.2). This indicates that the temperature at station 1 is controlled by the discharge of coolant water from the thermal power plant and heavy inflow due to tsunami might have carried this water to other stations thereby increasing their temperature. Fig.2 clearly depicts that all the stations except station 1 show a temperature increase of 4°C after tsunami.

pH has varied from 8.3 to 9.3 in December and from 8 to 8.7 in January (Fig.2). Lower pH during post-tsunami shows the dilution due to the invasion of the tsunami water.The lowest DO was observed at station 2 due to its close proximity to the industrial discharges. The observed variation between surface and bottom DO was negligible which proves the extensive mixing of canal water due to tsunami.

There was no significant difference in salinity between the surface and bottom suggesting the absence of stratification (Fig.2). The analyses of pre and post-tsunami salinity indicated an increase of 1 ppt in station 1, 2 ppt in station 2 and 4 ppt in stations 3,4 and 5 (Fig.2). Such higher salinty in stations 3,4&5 are attributed to the entry of tsunami waves through the northern inlet (5,Fig.1) which used to be permanently closed. The field study and the local enquiry also showed the thinned out mouth sand bar and the over topping of the tsunami waves on 26, December.

The maximum depth of the canal recorded prior to tsunami was 2.5m, where as there was a 50% decrease in depth after tsunami in all stations except 5 (Fig.3). This could be attributed to the deposition of sediments brought and dumped by the ferocious tsunami waves. Except station 1, in all the other stations post-tsunami SPM were higher than pre-tsunami. This shows that tsunami disturbed the bed material which increased the sediments in suspension. In station 2, SPM in the bottom, both during pre and post-tsunami, was higher because the industrial sludge settled at the bottom.

Total phosphate and Inorganic phosphorous showed identical trends. From station 1 to 5 a decreasing trend was observed, but the wastewater released from aquaculture farms may be responsible for the slight increase in phosphate concentration at station 4 (Fig.3). Higher phosphate level was observed at the bottom than the surfcace. Post-tsunami scenario showed lower phosphate concentration than the pre-tsunami because of the dilution of nutrients due to invasion of tsunami water.

Distribution of Nitrite, Nitrate and Total Nitrogen between the stations showed similar trends (Fig.3). Highest nitrogen compounds at Station 2 and 4 are due to the industrial discharges and aquaculture activities respectively. A slight increase in the bottom values were observed. Similar to other nutrients, nitrogen compounds also showed dilution after tsunami.

Chlorophyll a showed a higher concentration at the surface than the bottom due to decrease in light penetration (Fig.2). Irrespective of the observation periods, Chlorophyll a concentration is the lowest at station 2 because of the industrial pollution. But Chlorophyll a values during pre-tsunami was higher than the post-tusnami indicating the dominance of blue green algae before tsunami.

CONCLUSIONS

Industrialization coupled with the urbanization has lead to stress in the ecological system of Buckingham Canal at Ennore. December 26, 2004 tsunami

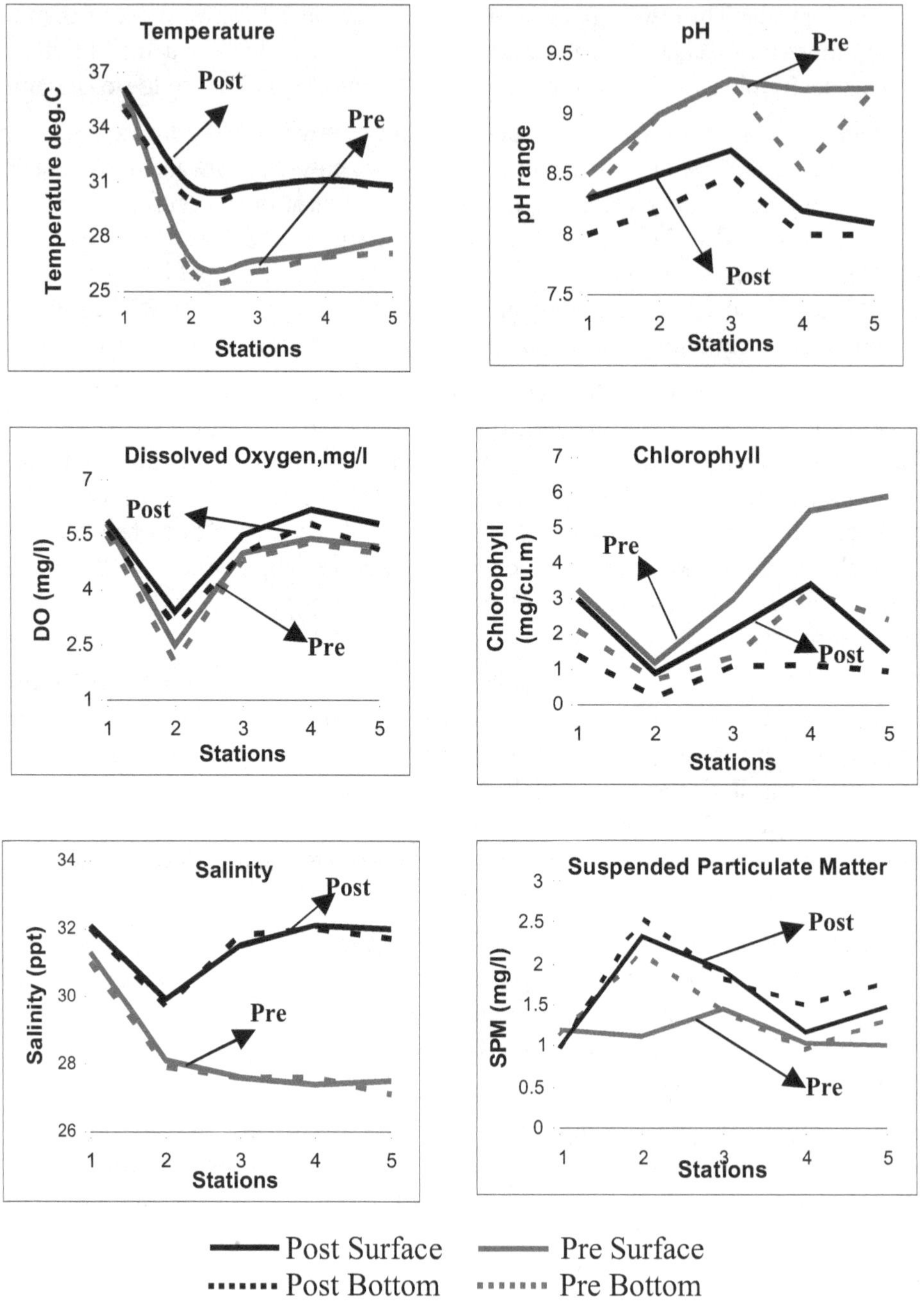

Figure 2: Physio - Chemical Parameters

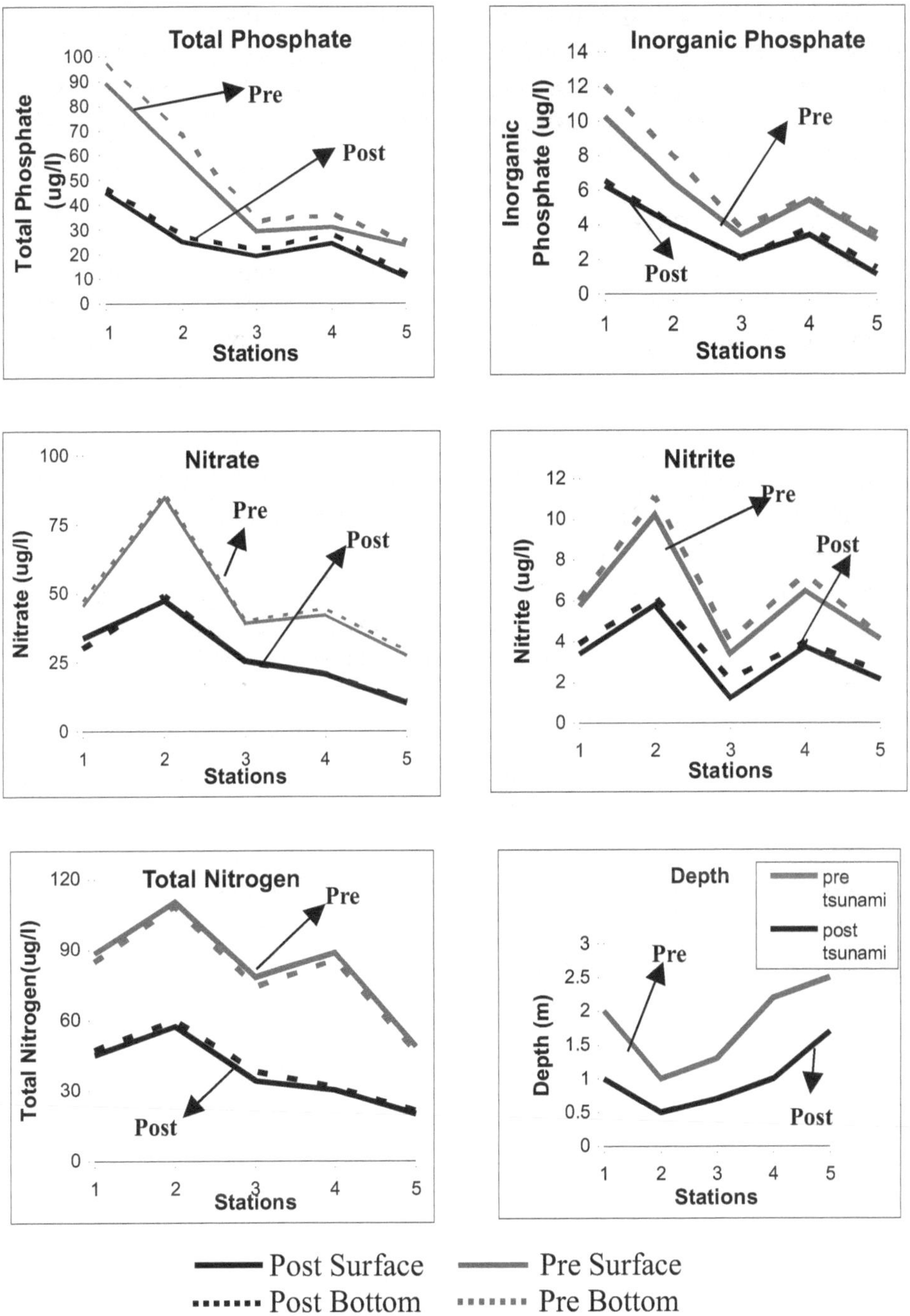

Figure 3: Physio-Chemical Parameters

brought devastating changes in the coastal stretch of Tamilnadu. Comparison of pre and post-tsunami water quality characteristics at Ennore, showed the incursion of sea water into the canal increased the water temperature (dispersing the coolant water from thermal power plant), salinity, SPM and reduced the nutrient levels, chlorophyll and depth.

ACKNOWLEDGEMENT

The authors gratefully acknowledge the financial support provided by the International Foundation for Science to carry out the project "Impacts of coastal pollution on biodiversity at Ennore, India" at Centre for Environmental Studies, Anna University, Chennai, India.

REFERENCES

Boesh, D. (1982) Ecosystems consequences of alternations of benthic community structure and function in the Network Bright region, Columbia S.C. Estuarine Research Federation.

Environmental Protection Agency (1981) Draft environmental Impact statement on the Hudson River PCB reclamation demonstration project prepared by WAPORA Inc. under the direction of EPA. Region II Personal Network: USEPA, Region II, 26 Federal plaza.

Grasshoff, K and Kremlig (1992) Methods of sea water analysis.

Levin, S.A. (1983) Mathematical modelling and the evaluation of the effects of anthropogenic stresses, R.Lambersson (ed), pp. 162-166.

Praveen Kumar, N., Desai, B.N. and Gajbhiye (1999) Comparitive account on zooplankton in polluted and unpolluted estaries of Malad creek. Indian Journal of Marine Sciences, v.16(3), pp. 281-291.

TSUNAMI –
RELIEF AND REHABILITATION

GEOMATICS IN TSUNAMI
S.M. Ramasamy, C.J. Kumanan, Brig. R. Sivakumar and Bhoop Singh (eds.), pp. 199-202
New India Publishing Agency, New Delhi (India)
E-mail : spjain_niph@rediffmail.com Web: www.bookfactoryindia.com

16

SYNDROME OF THE TSUNAMI (2004) RAVAGED SOCIETY - A MICROSCOPIC VIEW FROM NAGAPATTINAM, INDIA

C.Venkatachalam

Abstract

Tsunami (2004), that struck the Tamil Nadu coast has left deep marks in the minds of the affected society. The interview conducted in the coastal villages of the tsunami affected Nagapattinam has brought out many salient societal issues like, loss of leadership, widow hood, debt grip, psychological depression etc, which warrant deeper look to bring them out from the syndrome.

Keywords: Tsunami (2004), Social issues, Nagapattinam, India

INTRODUCTION

The damages caused by tsunami (2004) along the coastlines of most of the South Asian countries are indelible not only from the minds of the coastal population but also from the heart of the entire humanity. The worst affected in the disaster are the fishing community. Many of their mechanized and un-mechanized boats and catamarans were converted to useless debris and garbages by the heartless tsunami. When their settlements themselves were swallowed even with out chewing, nothing to talk about their fishing nets which were just blown off by the ferocious tsunami. In addition, the infrastructure facilities like roads, communication systems, water and power supply were also incalculably damaged. Next to fishermen community, the affected souls were the farmers whose cultivable lands were turned into marine wet lands with swimming and tossing fishes. The Indian subcontinent was facing such onslaughts of various natural disasters periodically but our morale always stood by our side prohibiting us declaring them as national calamities for getting international aids. Specifically in this case, the central and state governments, NGOs and Voluntary Organizations have risen phenomenally to help the affected who were in need. Despite best efforts taken to provide relief and rehabilitation to the affected people, the tsunami has seedled some societal issues / syndrome, which may take long years to settle; it may not settle too. The paper discusses some of such issues gulled out through the interview conducted amongst various cross sections of the affected community.

SOCIETAL ISSUES OF TSUNAMI

On the basis of the interview, the author could classify the issues in to different categories and some of them are dealt below.

Loss of Leadership

Mr. Ramu (29) is a fisherman engaged in fishing since his childhood. He lost his properties worth of Rs. 50,000/- and above. His thatched house has been washed away by the tidal waves. He could not resume his normal work even though the NGOs have provided the boat on installment basis. Besides, the Government too have provided marginal assistance. Apart from the vanished house, he lost his father too, which only has disturbed him to the maximum. Until tsunami, his father use to guide him, take the family chores and share the responsibilities of the family. His presence only was shaping the family. The sudden demise of his father unimaginably left a major vacuum over all, right from settling the future of his unmarried sister to psychologically building up his widowed mother and to carve out his future livelihood. The interview has revealed many such cases of leader / leadership lost families.

Widowhood

The 55 years old Mr. Natarajan lost his house and wife to the tsunami waves. At present he is living in a temporary settlement given by the NGO. The loss of his spouse has created a major imbalance in his life. He is unable to resume his normal work because he seems to have lost his confidence after his wife's death who used to give him moral support and stood by his sides in all activities. He feels that a widower's life is very difficult at his age. He appears to have lost all confidence and feels that without spouse he may not be able to manage the family chores cropping up from time to time. He obviously expressed his feeling that tsunami stopped most of his progress. Many such people were met during the interview. While this is the issue of the widowers, the widows having infant children have chains of problems, like their livelihood, bringing of the children and more the security for them etc.

Debt Grip

Mr. R. Mariappan's (45) major occupation was agriculture. He had two acres of land for his life until tsunami day. He could have harvested the paddy; without knowing the unfortunate happenings he had fixed the labourers for 27th December 2004 to harvest, but the tsunami tidal waves destroyed the crops. He now helplessly stare at his lands invaded by the sea water in the arable land and he feels that the land can not be used for few more years because of salinity. According to him the primary source of income is completely stopped, and at the same time the borrowed money from other sources are yet to be settled. He is moving around helplessly and tries to sort out his problems. He feels that he does not know jobs other than agriculture and there lies the whole trouble. The interview indicated many cases in such category; whether fishing or farming the poor people who borrowed lump sum amount for their livelihood. They are at loss now what to do with as they will have to settle the loan and then only their livelihood can begin by borrowing money again when the normalcy comes.

Psychological Depression

Mr. C. Raja, (40) is a fisherman who has lost his house and majority of the essential belongings. Lot of changes has been taken place in his life after tsunami. He is of the opinion that still more efforts would have been made to provide assistance to the real victims. He is psychologically disturbed very much due to unprecedented damages caused by the tidal waves.

Mr. P. Raja, (40) is another victim of tsunami. He lost his properties worth of Rs. 50,000/-and paddy field completely. A cyclone in 1995 had thrashed the properties and within a decade the tsunami created a vacuum in his life. The family is at present living with many hindrances. The helps came in the form of assistance could only partly save him and could not solve many of the tsunami created problems. Now he feels that whatever happening in and around him are beyond his control. He uncomfortably speaks about the social imbalances caused by tsunami.

25 years old Mr. Rehman is a fisherman and tells that the fishing job is the toughest one. It always threatens the life of fisherman folk. He expressed that the publicity and the real help by the agencies do not match. A couple of years ago he lost his spouse to viral fever. He has been in struggle with the children and the tsunami added some more troubles. He says that man's life consists of many ups and downs. He says that he cannot try for any other job, because he likes the fishing job even though it created lot of inconveniences after the tsunami. Such psychological depression and lack of confidence were observed to be major issue found in many tsunami affected regions.

PHILANTHROPY

Mr. Sethupathi Nattaar has narrowly escaped from tsunami at Aarukattutheru village of Vedaranniam taluk. In this village, his family has considerable amount of land and has in a position to control the fishing occupation too. Despite the damages created to his properties, fishing boats etc, he reluctantly accepted the compensation extended to him and to his family. He said that his family can manage the situation but the real victim should get all the available help on priority basis. His actions reflect the core culture of India. Even he never accepted the recompense for his children's education. He has said that the assistance should be *"help the people to help themselves"*. Interview has indicated the availability of few such persons too.

Will Power

Fishing was a major occupation for Mr. Irulappan and his family and on rental basis he was giving away the boats to other fishermen in his village. Since his childhood he knew fishing only. The tidal waves have destroyed most of his boats and nets but he appeared to be very strong enough to face any tasks fixed by the tsunami. According to him, fishermen lives are always challenging, tsunamis may come and go but the coastal people cannot leave the seashore. He spoke comfortably that the tsunami could create damages only to the materials, and not his might. The interview with him has revealed the will power deeply embedded in him.

DISCUSSION AND CONCLUSIONS

The case study is nothing but 'study about the single unit'; the Social Scientists in general and Sociologists in particular use this method to gull out the inner feelings of the concerned individual in qualitative methods. Frederic Le Play, Herbert Spencer and Berger are the leading scholars who used this particular method to draw the meaningful conclusions. The essential efforts have been taken in the field at Vedaranniam Taluk of Nagapattinam district to understand the problems of some victims on tsunami. When every individual is unique, the selected cases brought different paradigms of the tidal waves tsunami. The interview has brought out certain salient syndromes related to loss of leadership, widowhood, debt grip, psychological depression, etc. and this need deserving actions to bring them back to their normal cycle of life. The study is progress and these are some mid course observations.

GEOMATICS IN TSUNAMI
S.M. Ramasamy, C.J. Kumanan, Brig. R. Sivakumar and Bhoop Singh (eds.), pp. 203-207
New India Publishing Agency, New Delhi (India)
E-mail : spjain_niph@rediffmail.com Web: www.bookfactoryindia.com

17

TSUNAMI: THE CITIZEN'S PERSPECTIVE

T.Dhanapal

Abstract

During natural disasters, mostly panic and curiosity lead to greater causalities. But the tsunami is such a disaster which no one have witnessed during their life time nor heard of from their ancestors. On 26, December 2004, tsunami has struck all of a sudden many South Asian coasts including the east coast of Tamil Nadu and Andhra Pradesh causing greater devastation and causalities. This historical event has taught us many lessons and the paper deals with some aspects related to tsunami from the citizen's view point.

Keywords: Natural calamities, Tsunami mechanisms, Mitigation strategies, Citizen's perspective.

INTRODUCTION

The tsunami that ravaged the coasts of South India had left tell tale signs that affirm the need for early warning systems as well as well-planned mitigation measures for the benefit of the affected. Over the last 12 months, since the tsunami struck the east coast of India on 26 Dec. 2004, numerous seminars, discussions, programmes and project proposals have been deliberated on various aspects that deserve attention. Government agencies, self-help groups and non-governmental organizations (NGOs) have done commendable post-tsunami work. All sections of the society have responded both emotionally and rationally to the event in an unprecedented manner. In retrospect, to the informed citizen – one who has heard about the event, read about it and discussed it with others, though not have had the 'experience' of the natural disaster by himself / herself - what do all these mean ? the tragedy and its aftermath. This paper documents the issues that a citizen would like to ponder over so as to understand the various dimensions of the post-tsunami activities and draw a few lessons that the tragedy had taught us. The major issues pertain to (i) the mechanisms (ii) the coastal vulnerability and the need for assessment (iii) mitigation strategies (iv) media exposition (v) communication networking and (vi) educational facilitation.

MECHANISMS

Since the day tsunami struck, there have been many discussions on the nature of the calamitous event. Foremost, among the issues, one that comes to the mind of the citizen is how did the tsunami originate? Especially in a country like India where the event struck with no warning and where the word 'Tsunami' itself has never been heard of by the common populace, it is but natural that people would want to know about the physical mechanics of the phenomenon. The rude shock to which the people were subjected to compels them to query as to 'how this act of god' takes place. Undoubtedly it is the bounden duty of the 'knowledgeable persons' – here it is the Geo scientists and technologists who have knowledge of the dynamics of nature - to educate the common man in understanding the phenomenon. 'Understanding' is the key to 'appropriate response' to such events in future. The citizen looks forward to an explanation in an 'easy to understand' terminology. Scientific explanations rely on specialized terminologies; however, terminological complexities will have to be reduced to the minimum when explanations are provided to the common man. Undoubtedly, the task is a daunting one, with much of the population which had suffered the tragedy being illiterate or somewhat literate. Further, the reaction of a community, like the ones living in geographical areas, such as coasts, hills and valleys – areas which are prone to natural calamities – will certainly be characterized by 'fear'. Fear psychosis will have therefore to be considerably alleviated through explanations about these disasters, particularly on their origin and the precursor symptoms, if any. There must be sustained efforts towards creating adequate awareness on natural calamities, be it tsunami, earthquakes etc.

VULNERABILITY ASSESSMENT

It is imperative that vulnerability studies of the areas prone to natural disasters, should be undertaken by competent agencies, be it government or university. The unpredictability of natural calamities in fact warrants detailed scientific studies on the possible vulnerable geographical areas. The areas affected by the tsunami 2004 prima facie qualify for detailed and focused studies by the scientists / academicians from the concerned core disciplines; as at many times after such calamities "Everybody can do everything" philosophy prevails. The Govt and the funding systems need to ponder over this – which branch of science or which are the branches of sciences and in what prepositions and amalgamations can provide real scientific solutions – be it predictability / fore warning, vulnerability, short term mitigation, long term rehabilitation interms of society or environmental resources etc., get those brains coalesced and the findings are translated in to action. Actual science and technology charged remedial strategies are mandatory in these types of disasters; to that matter in any developmental planning.

However such studies need not be restricted to the genesis, vulnerability and preventive measures alone, equally important is the making of an inventory of infrastructural facilities already existing in the vulnerable areas, including residential and official premises and service facilities that have been situated in such areas. An estimate of the value of the facilities, including machinery – for instance those of off-shore refineries – will have to be made. Such documentation would lay bare the probable economic loss that might occur in the event of a natural calamity. Since maximum destruction is likely to be in specific areas like coastline, it would be prudent to make an advance evaluation of the facilities to make a fair assessment of economic loss that might result in the event of a calamity. An additional advantage of possessing such information would be easy identification of infrastructure which will have to be protected to avoid heavy economic loss. For instance, prompt action to protect electrical supply stations would ensure both limited economic loss and continuity of power supply to ensure relief operations. A cascading effect of continual power supply would be continual water supply, since power supply would ensure operation of pumping stations. Thus, precaution in the form of vulnerability studies is a much desired activity that should be undertaken with no further delay.

MITIGATION STRATEGIES

The devastation an event like a tsunami has demanded extraordinary efforts on the part of the Government machinery in dealing with the situation. The mechanisms to deal with such a situation are bound to be stretched to the fullest. Understandingly there are limitations, in such large scale operations, though by and large succor is provided to a larger population than envisaged, soon after the tragedy struck. The limitations of relief – operations, which at times get linked to bureaucratic accountability, stand foremost in the mind of the common man, particularly in the minds of the affected people. What ever extra care and committed efforts Government machinery takes during such crises, invariably the common man gets carried away by the short comings in the implementation of programmes announced for mitigating their suffering; due to their distress and state of mind during those crucial periods, the panic gripped affected section of the society comes to the psychological zeroness and totally depend on the others. Hence, such reactions, that are chiefly prompted by the personal loss of property, belongings and the resultant syndrome should be negotiated with humane consideration. So any disaster particularly of a high magnitude, such as tsunami, is bound to test the capability of the administrative structure. Thus the need of the hour is a comprehensive mitigation strategy taking into consideration the psychological trauma, the victims are likely to undergo. More than anything else, a coordinated and concerted effort would appease the emotionally charged populace rather than skewed attempts that aim to provide relief through random relief measures. One of the seminal post-tsunami activities would be livelihood protection to the affected population. The relief and rehabilitation strategies adopted by the Government and NGOs during the tsunami disaster in worst hit areas of Tamil Nadu need appreciation. But this can be pier reviewed, for short comings and a comprehensive Tsunami mitigation package imbibing the strong S & T inputs may be rewarding and reduce causalities, loss and our budget too.

MEDIA EXPOSITION

Media have been the main source of data and information to the common man – the citizen – who is 'so near yet so far' from the calamitous event. Visuals, both in printed and televised format, evoke sentiments and sympathy for the affected. News paper reports and editorials bring to one's knowledge the extent of the damage done as well as the measures being taken to address the sufferings of the affected people. The actions of 'the appropriate response system' are presented to the public. Discussions focus also on how man's assault on nature might trigger calamitous events; tampering with nature – both flora and fauna - might result in extraordinary events. Compliance with environmental security might ensure a comparatively lesser damage to lives and property. It is thus the media which conveys to the public that though no one can stop such calamitous events, at least steps can be taken to be prepared for such events. The media, thus, have to constantly strive to assure the public that the Geo scientists are closing in on an understanding of the dynamics of such unpredictable events.

COMMUNICATION NETWORKING

Information access is crucial to prompt action by the relief machinery. System failures often are the result of communication breakdown between and amongst the departments or organizations that deal with the situation. The absence of information at one level leads to inaction at another level. It is time a comprehensive communication network is established as a measure of preparedness for meeting events such as tsunamis, which strike unannounced. It is worth remembering that occasions such as natural disasters create panic among victims. Establishment of a proper communication network which would ensure information flow, in both directions, from the field officials to the administration and vice-versa, will give the necessary semblance of order and to help control the panic stricken victims.

EDUCATIONAL FACILITATION

Curiosity is inherent to human beings. Natural calamities, particularly tsunamis, kindle curiosity. However, when curious events take place the dictum of 'safety first' suffers. There have been reports of people who perished because they curiously went down the beach to know what was happening. Post-tsunami relief operations cite instances of children suffering from serious illnesses resulting from something as simple as taking a drink of contaminated water and of those, particularly from the fisherman community, suffering from illnesses due to lack of proper hygiene which aggrevate under conditions of deprivation. Such instances of the lack of knowledge of safety measures are reflective of the need for education. Proper response to a calamitous event depends much on knowledge of what to do and what not to do. It is therefore essential that steps are taken to educate the people, particularly those residing in disaster prone areas, on 'behavioural patterns' when eventualities take place so as to facilitate them to handle the situation effectively.

CONCLUSION

In conclusion, the citizen, the common man looks forward to an integrated approach based on understanding of both the 'event' and the people who are victims of the event. A comprehensive approach that considers the six issues referred above would go a long way in establishing a system to deal with natural disasters effectively. Such a system, incorporating scientific management and the humane aspects, promises a ray of hope for all of us who have pity for the victims and fear for ourselves against such natural occurrences.

GEOMATICS IN TSUNAMI
S.M. Ramasamy, C.J. Kumanan, Brig. R. Sivakumar and Bhoop Singh (eds.), pp. 209-213
New India Publishing Agency, New Delhi (India)
E-mail : spjain_niph@rediffmail.com Web: www.bookfactoryindia.com

18

PARTNERS IN ACTION: NGOs IN TSUNAMI RELIEF AND REHAB.

Annie George

Abstract

The paper deals with the roles played by the NGO's in the relief and rescue phase of Dec, 2004 tsunami and the effective role played by NCRC -NGO Coordination and Research Centre - in the rehabilitation phase along Tamil Nadu coast by creating successful partnership with government and the public.

Keywords: The Role of NGO's, Tsunami 2004, Relief and Rehabilitation, NGO Coordination Research Centre.

INTRODUCTION

The scale of the tsunami disaster was unprecedented and the response too was out of the ordinary. Tamil Nadu was the worst affected state in India with Nagapattinam district being the most affected with over 6000 lives lost, about 40,000 houses damaged or destroyed and 1.96 lakh people affected. Relief poured in from both government and non-governmental organizations, including the religious organizations, and the first few days witnessed heartwarming outpourings of help and support.

There were about 400 NGOs and an equal number of volunteers to help in the relief operations which included food, clothing, household utensils and bedding. With such a large scale of external support, mainly from organizations and people from outside Nagapattinam, it was not surprising that chaos and confusion marked the first few days of relief operations still. Based on the experience of some professionals experienced in disaster management, a coordination platform was created called the NGO Coordination Cell. This platform was managed by two NGOs, who had long standing field presence in Nagapattinam, viz., South Indian Federation of Fishermen Societies and SNEHA, with the mandate and support of the District Administration. The main role of this platform was to meet the field based requirements of all the relief camps with the resources and materials available with the NGOs and the Govt. The paper briefly deals with challenges faced by the NGO's in relief and Rehabilitation of the historical Dec 2004, Tsunami.

PUBLIC–PRIVATE PARTNERSHIPS IN RELIEF AND THE REHABILITATION

The Tamil Nadu government took the lead in rescue and relief operations with the disaster management and mitigation department of the Revenue Administration section working through the District Collectors. The government infrastructure is systematic, far- reaching and centralized. Despite the fact that top officers were sent to the field to oversee rescue and relief, there were areas where the NGOs were able to respond little faster – their decentralized set up meant that their response time was quicker – an absolute must in cases where time is critical. Also, being tuned into the community's immediate and actual requirements, they were able to respond to specific needs– whether it was food, utensils or hygiene kits. Their need- based response also did not require identification papers or ration cards to qualify for relief. NGOs relationships with donor agencies meant that funds were made quickly available; their relationships with corporate entities meant that sourcing of items such as large- scale requirements of disinfection agents was faster as was the mobilizing of debris removing equipment such as bulldozers and earthmovers. Similarly, some NGOs, because of their previous contacts and work with certain communities, were able attend to their needs. In general, it was easier for the administration to work with the NGOs for rapid response and to ensure that the maximum numbers of affected people receive due relief.

And, this is what set apart the relief and rehabilitation efforts in Nagapattinam- this effective "public- private partnerships" between the Govt. and the NGOs. Each had its own inherent strengths and weaknesses but together, there was a synergic relationship that built upon each other's strengths and mutually supported each other.

With the evolution of the relief phase into the rehabilitation phase, the number of NGOs in Nagapattinam dropped to about 190 and the individual volunteers left to go back to their normal lives. The 190 NGOs that remained took on long- term activities with singe- focus areas rather than multiple foci, largely based on their areas of expertise. There are presently about 45 NGOs working on permanent shelter construction, about 23 NGOs involved in reclamation of affected agricultural land, about 40 NGOs concentrating on children related activities and the like.

The NGO Coordination Centre continued its existence into the rehabilitation phase as the NGO Coordination and Resource Centre (NCRC) with its main areas of focus being:

- Information Dissemination,
- Sectoral Support and
- Advocacy

The NCRC strongly subscribes to the belief that access to information is one of the key factors in ensuring equitable distribution of the compensation packages and supportive measures initiated by the Govt. To this end, the NCRC has set up 13 Village Information Centers across the coastline of Nagapattinam district.

Most of the NGOs supporting rehabilitation have the commitment and the resources but not necessarily the expertise required to deal with the multi- faceted issues that the tsunami devastation has given rise to. It was infeasible to have the necessary skill- sets, cutting across all these sectors, within the same group and this necessitated building up and sharing of sector-based support systems. The NCRC has been actively involved in bringing together NGOs working in these select sectors, providing a wider platform for the discussion and dissemination of perspectives, policies and best practices as well as trying to build a consensus on the best approaches within the larger framework of replicability, long term sustainability and overall development. The Sector Support Unit serves as an information and knowledge centre for each sector identified, which would also keep pace with the developments in the respective sectors.

With a destruction of this magnitude, quick responses by the policy makers, although extremely necessary, would impact the various strata of the affected in different ways. The NCRC, with the knowledge of the ground realities as assessed by its Village Facilitation Units and the benefit of expert advice from acknowledged authorities, is in a key position to assess the effectiveness of the policies vis- a- vis equity, access and actual availability at the grass root levels etc. The NCRC utilizes this knowledge to flag the issues and concerns, emerging from the field level, to the notice of the policy makers, thereby trying to ensure that:

a. All communities and population, including the vulnerable, have equal and free access to the relief and rehabilitation packages and process and are integrated into the rehabilitation and development processes
b. The policies are in keeping with the diverse needs and requirements expressed by the various strata of the affected
c. The approaches of the various players are synergic and in keeping with the overall developmental framework envisaged

NCRC has been actively involved in advocacy issues in the sectors of fisheries, shelter, agriculture and education. The inputs provided by the NCRC, in consultation with the NGOs, communities and experts, have been, by and large, accepted and translated into action.

The last year has witnessed rapid changes in all sectors. Fisheries livelihood had been resumed as early as June with NGOs very generously contributing boats and fishing gears. The distribution of boats has far surpassed those lost and the impact of such largess yet needs to be studied. However, early assessments show that the market forces are taking care of this and the boat movement has stabilized to just what the system can comfortably hold. Even in the next most important livelihood, agriculture is concerned, despite the fears that no sowing could be done for the next couple of years till the land is completely desalinated, more than 70% of the affected land was desalinated and sowing done. The reclamation process identified was a mix of the Department's package and the experience in organic farming practices followed.

About 17,500 permanent houses are being constructed by the NGOs, the largest permanent shelter construction in Tamil Nadu. These have been designed based on the requirements expressed by the community on to lands provided by the Govt. Identification and purchase / acquisition of land in Nagapattinam has been complicated by the facts of low- lying terrain, need for the land to have sea access for the fishermen community as well as the overriding concern for safety. Despite a hefty requirement of 900 acres fulfilling all these parameters, the Govt. has been able to identify, procure and allocate about 85% of this. About 2000 houses will be complete by end Dec. 2005, and 14,000 are in various stages of construction.

The common thread running through all these rehabilitation activities have been the continued relationship that is being maintained between the NGOs and the District Administration. While the District Administration is in charge of the broad policies and framing the guidelines of rehabilitation efforts, they are also very open to the field based concerns and requirements as brought to their notice by the NGOs. As mentioned earlier, with an effort of this magnitude, it is not uncommon that there would be communities or individuals overlooked. However, this oversight is immediately corrected if raised by the NGOs. Similarly, the NGOs are guided and helped by the District Administration on the overall policies and mandatory requirements like sanctions, approvals and exemptions.

This is not to say that there have not been differences of opinions and clash of priorities. However, these have been solved through discussions, debates, bringing in of technical expertise and fueled by the single agenda of doing what is best for the community. A free and open relationship enables and enriches rather than disables and constricts. This has helped both the Administration, the NGOs and more important, the affected community in rebuilding shattered lives, livelihoods and a new vibrant Nagapattinam.

CONCLUSION

It appears a nearly ideal mix – the blending of the short term quick and appropriate response by the non governmental organizations coupled with the wider coverage and sustained activities by the government – in responding to a disaster of the magnitude of the tsunami. Perhaps the most important outcome of the tsunami disaster is the fact that governments and non governmental organizations can work in partnership and that too, successfully. As we approach the first anniversary of the tsunami disaster, it is time to recapitulate, to see where the response was right and what was left out where, so that the next time around, the response could be even better.

GEOMATICS IN TSUNAMI
S.M. Ramasamy, C.J. Kumanan, Brig. R. Sivakumar and Bhoop Singh (eds.), pp. 215-220
New India Publishing Agency, New Delhi (India)
E-mail : spjain_niph@rediffmail.com Web: www.bookfactoryindia.com

19

THE TSUNAMI DISASTER AND TAMIL NADU'S RESPONSE, INDIA

R. Santhanam IAS

Abstract

Though people belonging to different disciplines from the common citizen to top level administrators are familiar with cyclones, floods, earthquakes, landslides, droughts, etc., the tsunami which struck and devastated over 1000 km long Tamil Nadu coast was not only unexpected but also an unknown phenomenon for the people of this part of the region. The Government of Tamil Nadu, though faced with this disaster for the first time, has efficiently managed the crisis owing to clear policy directions from the top political executive of the government, the planned working of the ministers, the committed service rendered by the senior officials and the tireless efforts of district collectors and their supporting staff down the line. The immediate measures taken in search, rescue and relief operations are detailed in this paper. Also covered are the rehabilitation measures undertaken including those for orphaned children and unmarried girls and the steps adopted for the prevention of epidemics.

Keywords: Tsunami Disaster, State Administration, Relief and Rehabilitation, Tamil Nadu.

INTRODUCTION

India is traditionally vulnerable to natural disasters on account of its geo-climatic conditions. Floods, droughts, cyclones, earthquakes and landslides have been occurring regularly in some part of the country or the other. In Tamil Nadu, the unprecedented drought in 2002-03, followed by another drought of comparatively less severity in 2003-04 and the floods in the northern coastal districts in October-November, 2004 had kept the State machinery on constant alert. But the tsunami that struck the coastal districts on 26th December, 2004 was an altogether new and nightmarish experience. It was a disaster, the like of which, had not been seen before. The suddenness and the ferocity with which the waves gushed into the habitations along the 1076 km long coast line, taking the people totally unaware, caused unprecedented damage, death and destruction and posed the greatest challenge to the State administration. However, the state faced the challenge successfully, thanks to the vision and commitment shown by the Government and the timely help rendered by a number of non-governmental organisations and groups.

STATE'S DISASTER PREPAREDNESS

A brief review of the state of preparedness to tackle the disasters would be relevant before one gets into the details of how the crisis was managed. Disaster Management assumed a new meaning and urgency in India following the three major disasters that occurred in different parts of the country between 1993 and 2001 viz., the Latur earthquake in 1993, the Orissa super cyclone in 1999 and the Bhuj (Gujarat) earthquake in January 2001. All these disasters had caused widespread damage to life and property. The observance of the decade 1999-2000 as the International Decade for Natural Disaster Reduction and the emergence of television as a powerful medium of communication contributed to bringing the need for disaster management into sharper focus. The State of Tamil Nadu has made great strides since 2003 in spreading awareness on disaster management and building up an effective system for tackling disasters. A Disaster Management Policy which focuses more on prevention, pre-disaster preparedness and mitigation rather than post-disaster activities of relief, rehabilitation, restoration and reconstruction has been put in place. A Disaster Management Authority with the Chief Secretary of the State as the Chairperson has been constituted to advise the Government on all policy issues relating to disaster management. The Department of Revenue Administration, Disaster Management & Mitigation is now implementing a Disaster Risk Management Programme (DRMP) in collaboration with UNDP and the Union Home Ministry in six multi-hazard prone districts. This programme seeks not only to generate awareness among all stakeholders about disaster mitigation, but also lays emphasis on the involvement of local community in the preparation and execution of Disaster Risk Management Plans at the village, panchayat, block and district levels. The districts worst affected by the tsunami viz., Nagapattinam, Kanniyakumari and Cuddalore, were all implementing the Disaster Risk Management Programme and the community in many of the villages in these districts had already been sensitized about the need for preparedness. Tamil Nadu also has a Disaster Mitigation Plan to tackle the situation arising out of floods, cyclones and droughts which have been occurring in the State periodically. On the basis of the experience gained in handling these disasters, Government have codified the preventive measures that have to be taken well before the commencement of the Northeast Monsoon. Contingency Plans have also been prepared for all the districts and regular protective and preventive measures are initiated well ahead of the onset of the monsoons.

It is apparent therefore that the State machinery has the capacity and the wherewithal to tackle unexpected disasters. The tsunami however was altogether a new phenomenon which took everyone to surprise. In the absence of a tsunami warning system, the evacuation of the people living in the vulnerable areas could not be undertaken. But once the tsunami had hit, the system so assiduously built up over the years, ensured that rescue and relief operations were mounted on a warfooting.

MISSION MODE DISASTER MANAGEMENT

The giant tsunami tidal waves which were triggered by an earthquake of 9M off the west coast of northern Sumatra and an earthquake of 7.3M in Great Nicobar, hit the Tamil Nadu coast between 0830 hrs and 0900 hrs and by 1030 hrs senior officials of the government dealing with disaster management and response assembled at the Hon'ble Chief Minister's residence. An assessment of the situation was made based on the reports coming from the districts. On the 26th afternoon, the Hon'ble Chief Minister took an aerial survey of the coast. She visited the three districts that suffered the maximum damage viz., Nagapattinam, Kanniyakumari and Cuddalore on the 27th and consoled the families of tsunami victims. Thanks to the clear directions from the political executive, the Government machinery moved into top gear. The system of monitoring developments and attending to immediate needs was activated. The highest priority was given to search and rescue operations. The Army, the Navy, the Air Force and the coast guard were alerted and pressed into service on the 26th within hours of the tragedy. A year ago, the Tamil Nadu Fire and Rescue Services had raised a Search and Rescue Team in each of the 30 districts in the State, each team comprising 20 commandos and 4 Station Officers and headed by an Assistant Divisional Officer. These teams which had been given regular training were rushed to the affected areas from the neighbouring districts. The State Transport Corporation ran a number of services to move the affected persons to safer places.

The Hon'ble Chief Minister deputed a number of Ministers and Chairpersons of the Public Sector undertakings to Nagapattinam, Cuddalore and Kanniyakumari districts to guide and oversee the relief operations. She also deputed a number of senior officers to assist the Collectors in relief work. Officials at different levels from Revenue, Rural Development, Public Health and Public Works Departments were mobilized and sent to the affected districts to help the district administration to restore normalcy quickly. Sanitary workers in large numbers from the Corporations of Coimbatore, Madurai and Tirunelveli and the adjoining municipalities were rushed to the affected districts on 27th December itself. Earthmovers and similar heavy machineries were mobilised from Government departments and private sources and sent to the affected districts to clear the debris and enable officials engaged in relief works to gain access to the habitations. Multi-disciplinary teams, each team comprising a Minister, a senior I.A.S. official, officials from Revenue, Rural Development, Public Works, Water Supply and Public Health Department, were put in charge of 7 to 8 villages. These teams reached every hamlet that was affected and ensured that search and rescue operations were carried out. Nearly half a million people had to be evacuated and 412 relief centres were opened in schools, colleges and marriage halls to accommodate them. The fact that educational institutions had been closed for Christmas and the marriage season was to commence only in the middle of January helped in opening the relief centres immediately.

The institution of the District Collector always plays an important role in the administrative structure of the State and the Collector assumes the role of a crisis manager in any disaster situation. During the tsunami disaster, the special empowerment of the district administration has paid great dividends in the rescue and rehabilitation mission. A crucial decision to place sufficient funds at the disposal of Collectors and empower them to spend on immediate needs helped them to clear debris, remove bodies and make arrangements to cremate or bury them as per community custom. The bodies were photographed / videographed before they were handed over to the relatives and the legal requirement of post mortem in such cases was waived. Disinfectants like bleaching powder, slaked lime and phenyl were dispatched in large quantities. As a result of these measures almost all the bodies had been retrieved and buried or cremated by 31st December. One major achievement was that there was no outbreak of epidemics in the affected districts.

Relief poured in from all sources – State Governments, National and International NGOs, religious groups and thousands of individuals across the country who were moved by what they saw on television. The widespread media coverage had a flip side to it. It prompted the donation of inappropriate goods, especially used clothes, which were not accepted by most of the affected families. Special arrangements were made to receive the relief materials that arrived by road, rail and air and dispatch them to the affected districts. Coordination cells were set up at district and State levels to help the NGOs who had arrived in large numbers. Control rooms were opened at the Relief Commissioner's office and at the Collectorates to disseminate information to the public. The State Government created an exclusive section in its website on tsunami rescue and relief operations. The large numbers of dignitaries from the Central, other State Governments and also from foreign countries were briefed about the action taken by the Government. The electronic and print media were also appraised of the developments on a regular basis. Collectors were asked to deploy special teams to assess the damage caused by the tsunami. After a quick assessment, a detailed Memorandum was sent to Government of India on 3rd

January, 2005, seeking Rs.4800 cores as assistance. Government of India apart from sending a Central team of officials, sent a Joint Assessment Mission (JAM) comprising World Bank, Asian Development Bank and United Nation Organizations to estimate the losses and damages due to the tsunami. The Mission assessed the damages and losses at US $ 838.32 million.

In the days following the disaster, the State Level Committee on Calamity Relief Fund met almost three times a week to review the relief measures. Some of the decisions taken by the Government in the first week included sanction of immediate relief of Rs.1 lakh per person dead to the families affected, a well thought out sustenance package for about 1.2 lakh families whose houses were damaged, a separate package for another 1.8 lakh families who lost their livelihood, cash relief to those who had lost their animals and also to those who had sustained grievous injuries, etc. The Government also sanctioned Rs.40 crores for construction of temporary shelters to accommodate those who have lost their houses completely.

REHABILITATION PACKAGE

Once the relief phase was completed within a period of two to three weeks rehabilitation was taken up. Separate rehabilitation packages were announced for all the categories of persons who were affected - fisher people who had lost their boats and nets, agriculturists who had lost their crops or whose lands had become saline because of incursion of sea water, small businessmen and petty traders who suffered loss of property and income, students who lost their books, note books, educational certificates, etc., and all other similarly affected categories who needed support.

The most heartbreaking aspect of this disaster is the plight of children who were orphaned. While more than 200 children lost both parents, more than twice that number had lost either the father or the mother. Protecting these children from exploitation has been given very high priority. Though a large number of people, from India and abroad, offered to adopt these children, the Government decided to open orphanages and service homes for them with all amenities. Three orphanages and two service homes have been opened, the former to take care of 100 children in each and the later to look after the interests of orphaned adolescent and also unmarried girls. It has been planned to impart vocational training in these service homes to enable the girls to take up self-employment in future. The Government has also decided to open a fixed deposit of Rs.5 lakh in a bank in the name of each orphaned child/adolescent girl to be given to them with interest when he/she becomes a major. These children are also being provided psycho-social care by trained experts and volunteers. Equally poignant is the plight of parents who lost their children to the tsunami waves. Special programmes are being implemented to provide psycho-social care and rehabilitation of this group.

The Government has encouraged NGOs and CSOs (Community Service Organisation) to work in the affected areas. NGO co-ordination cells have been opened in the districts to facilitate their functioning. Separate schemes have been announced to encourage NGOs participation in long term rehabilitation / reconstruction programmes.

As a follow up of the appraisal by the Joint Assessment Mission, detailed negotiations were held with the World Bank and Asian Development Bank. Two major projects, viz., the World Bank aided Tsunami Emergency Reconstruction Project (TERP) and the Asian Development Bank aided Tsunami Emergency Assistance Project (TEAP) have been launched. When these projects are completed in December 2007, there will be a total transformation in the infrastructure and economy of the affected areas. A long term project, assisted by the International Fund for Agricultural Development (IFAD), will focus on the livelihood restoration of a large number of persons who were indirectly affected.

A study conducted by Fritz Institute of NGOs on tsunami affected families in India and Sri Lanka has revealed several features, two of which have a bearing on Government's response: (i) eighty six per cent of the affected families surveyed in India said that they received help from the Government in the first 48 hours and (ii) the affected people ranked the Government as the number one provider of aid on all the major dimensions of relief services. A strong leadership and a committed bureaucracy have made this possible. But disaster managers in Tamil Nadu are also conscious that along with Government's efforts local preparedness, participation and involvement are vital for effective relief.

CONCLUSION

Though, the tsunami was a totally unexpected disaster, the manner in which the policy makers and the state administrative machinery have handled it received appreciation from around the world.

TSUNAMI –
FJORD PROVINCES OF NORWAY

GEOMATICS IN TSUNAMI
S.M. Ramasamy, C.J. Kumanan, Brig. R. Sivakumar and Bhoop Singh (eds.), pp. 221-225
New India Publishing Agency, New Delhi (India)
E-mail : spjain_niph@rediffmail.com Web: www.bookfactoryindia.com

20

LANDSLIDES TRIGGERED TSUNAMIS IN NORWAY

SM. Ramasamy, Bhoop Singh and Brig. R .Sivakumar

Abstract

Subsequent to December 26th 2004 historical Tsunami, The Government of India has conceived a master plan for setting up Tsunami Warning System (TWS) similar to Pacific Ocean Tsunami Warning System. Concurrently, the union Govt has also initiated scientific studies through Department of Science and Technology and the significant amongst these are the

- *All India Co-ordinated Programme on Tsunami disaster and risk assessment using GIS technology and*
- *Bilateral programme with Norway to know the know – how and capacity building in the fields of Landslides and Tsunami*

As a part of this bilateral mission, the authors visited institutions connected to Geohazards and Tsunami affected areas of Norway. The article briefly narrates the salient inferences on the Tsunami scenario and the related studies in Norway.

Keywords: Landslides triggered tsunamis, Norway

INTRODUCTION

The 26th December 2004 "Asian Mega Tsunami" has opened up a new chapter in the disaster scenario of South Asian territorial nations including India. Most of these affected Asian countries, namely Indonesia, Andaman, Sri Lanka and India were not having adequate preparedness for Tsunami as it is first of its kind and these countries have not faced this before. Alarmed by this disaster, the Government of India has immediately swung into action to enter into the global network of Tsunami Warning Systems (TWS) analogous to Pacific Ocean countries. Concurrently, through The Department of Science and Technology, the Government of India, has embarked into two major programmes;

- ➢ Creation of large scale GIS data bases on the tsunami damages, risks and possible strategies for mitigation using Remote Sensing, GIS and ground surveys, to facilitate the concerned district administration, and
- ➢ Launching of bilateral collaborative programme with Norway to know the, know – how on Geohazards (landslides and Tsunamis), capacity building, exchange of scientists between these two countries etc.

As a part of such bilateral programme, the third author has visited Norway earlier and, the first two authors visited Norway later for a week during 5 – 11 September 2005. They have participated in the International conference on landslides and avalanches and also visited the West coast of tsunami gripped Norway. The paper briefly narrates the salient inferences made on the visit.

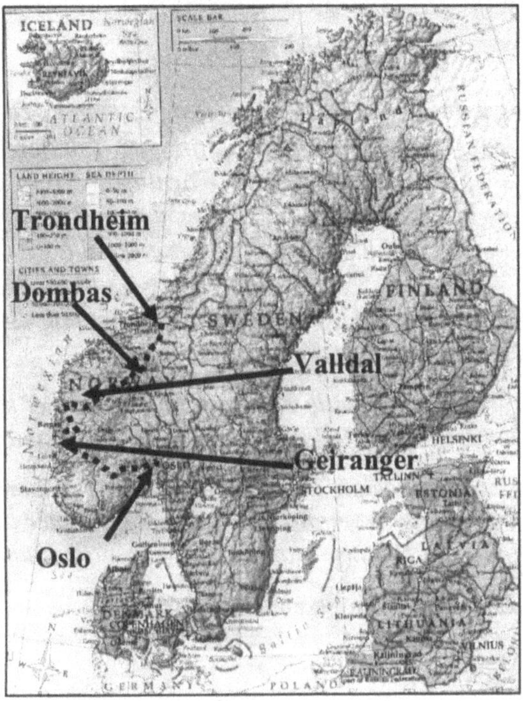

Figure 1: Showing the field trip routes in Western Norway.

GEOLOGY AND GEOLOGICAL PROCESSES

The NNE – SSW trending Norway peninsula exposes rock types of all ages of Arachaeozoic to Cenozoic era. The Arachaeozoic – Paleozoic piles of rocks form general NE – SW trend with faults dominantly aligned in NE – SW and NW – SE directions. The Norway peninsula was (Fig.1) capped by intense glaciation from 100,000 years to 10,000 years. Around 10,000 years, extensive deglaciation occurred, causing steep vertical escarpments and U shaped valleys along the old faults. Due to such deglaciation induced elastic rebound, the Norway peninsula was domed up with NE – SW axis forming sheet joints along with further deepening of U shaped valleys and steep escarpments on either sides of the valleys. Norway and its counter part in the peninsula, Sweden, has undergone an upliftment of 600mts in the eastern Sweden part and 200mts in the western Norwegian part of the peninsula thereby the hinge located along the West coast of Norway. Such tectonic activity and the resultant escarpments and the U shaped valleys along the west coast lead to the formation of Fjords etching the whole west coast of Norway. These Fjords are thus bounded on either sides by highly jointed (Fig.2&4) fractured and faulted steep escarpments with huge column of sea water in the valleys.

Figure 2: Glacial Escarpment and fracture controlled snow derived falls in Valldal – Geiranger Fjords

LANDSLIDES INDUCED TSUNAMIS

The Fjords as are flanked on either sides by steep escarpments with high degree of jointing, fracturing and the faulting due to still ongoing tectonic activities, the huge volume of rock mass often fall into the Fjords causing tsunamis. The authors have undertaken the cruise in one of the major Fjords in between Valldal and Geiranger (Fig.3).

The Valldal being located almost in the tail of the Fjord, has been witnessing tsunamis triggered by landslides more frequently. On either sides of the Fjords huge rocky escarpments are seen with active faults, lots of sheet joints and fractures (Fig.4, 5 &6) The N.G.I (Norway Geotechnical Institute), I.G.C (International Geohazards Centre) and N.G.U (Geological Survey of Norway) are carrying out detailed geotectonical investigations both along the escarpments and the Fjords floors for monitoring rock movements and to evaluate the past avalanches.

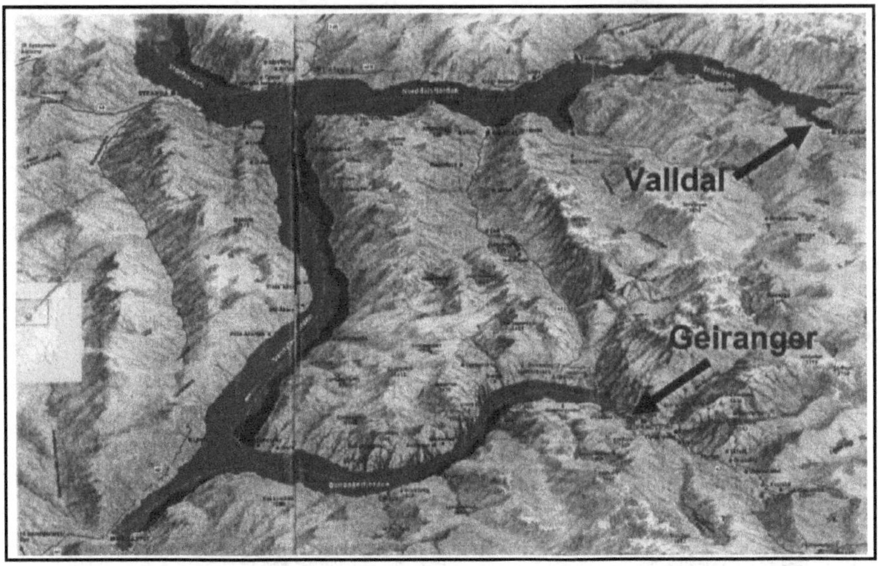

Figure 3: Valldal – Geiranger Fjord, Western Norway along which the authors have undertaken cruise from the former to the latter locations

Figure 4: Site of Investigation for Landslides (A & B), Faults / Fracture opening in glacial escarpment –Near Aknes (C) (Valldal –Geiranger Fjords)

ON GOING STUDIES IN NORWAY

The above agencies are carrying out studies in the Valldal – Geiranger Fjords to understand the Landslides triggered Tsunami phenomena for evolving protective, preventive and predictive models. These include

- Ground based SAR and Lidar interferrometry for detailed mapping and monitoring of the active slopes and escarpments which show signs of movements (Fig. 4a, b & c).
- Differential DEM based 3D modeling of avalanches and Landslides in Fjord provinces.

- Avalanche sediment mapping and DEM based sea floor / Fjord floor mapping to bring out the pattern, spatial distribution and the ages of the avalanche sediments in valldal – Geiranger Fjord floor to predict future possible events (Fig.5).
- The modeling and visualization of tsunamis (Fig.6).
- Tsunami risk mapping etc.

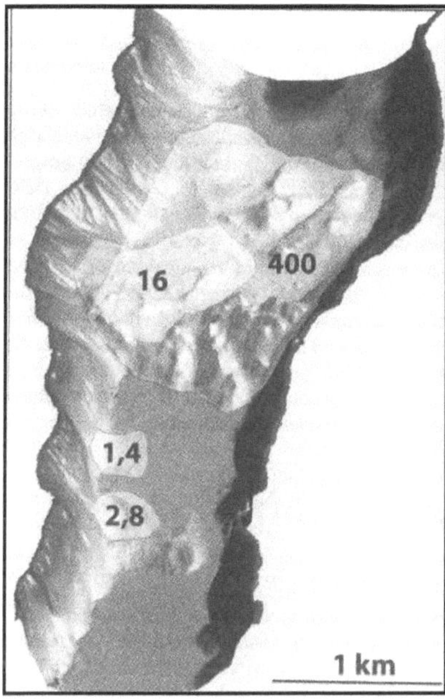

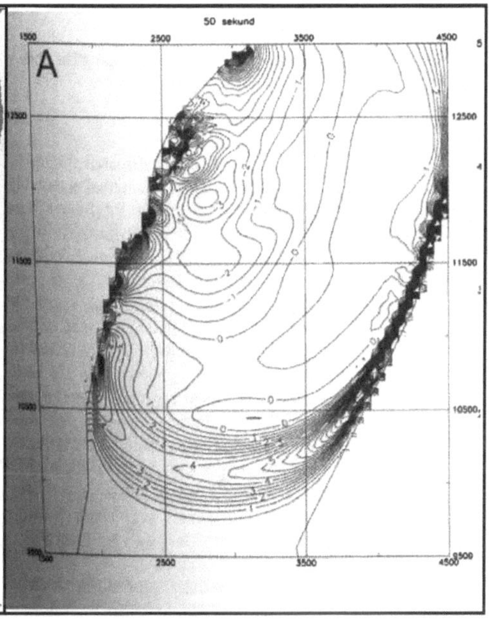

(Storfjorden, Western Norway)

Figure 5: DEM of Fjord floor

Figure 6: Visualisation of Tsunami Propagation in the fjord

These studies are expected to bring out a comprehensive picture on the Landslides triggered Tsunami disasters in Fjord provinces of Western Norway.

CONCLUSION

This short write up is a bird's eye view on the Tsunami related studies in Norway and the authors are grateful to the Scientists of N.G.U, I.G.C and N.G.I of Norway, for taking the authors to the field and provided adequate information. Fig. 5 & 6 were inherited from the publications of Norwegian scientists for whom the authors are thankful.

GEOMATICS IN TSUNAMI
S.M. Ramasamy, C.J. Kumanan, Brig. R. Sivakumar and Bhoop Singh (eds.), pp. 227
New India Publishing Agency, New Delhi (India)
E-mail : spjain_niph@rediffmail.com Web: www.bookfactoryindia.com

Author Index
(A-Z)

FSC
www.fsc.org
MIX
Papier | Fördert gute Waldnutzung
FSC® C083411

Zeitfracht Medien GmbH
Ferdinand-Jühlke-Straße 7
99095 Erfurt, Deutschland
produktsicherheit@kolibri360.de